6/30/93

# ATOMS AND QUANTA

Miraculous, the flight of measured thought
Crosses the rebel fire of burning youth;

Vernon Watkins
The Turning of the Stars

# ATOMS AND QUANTA

**DAPHNE F. JACKSON**

Professor of Physics, University of Surrey
Guildford, Surrey, UK

## SURREY UNIVERSITY PRESS

*In association with Academic Press*
*Harcourt Brace Jovanovich Publishers*
London   San Diego   New York   Berkeley   Boston
Sydney   Toronto   Tokyo

Surrey University Press
published by
ACADEMIC PRESS LIMITED
24-28 Oval Road,
London NW1 7DX

*United States Edition published by*
ACADEMIC PRESS INC.
San Diego, CA 92101

© 1989 Surrey University Press

This book is printed on acid-free paper ∞

**British Library Cataloguing in Publication Data**

Jackson, D.F.
  Atoms and quanta.
  1. Nuclear physics
  I. Title
  539.7

  ISBN 0–12–379075–1

Typeset by Colset Private Limited, Singapore
and printed in Great Britain by T.J. Press
(Padstow) Ltd, Padstow, Cornwall.

# Contents

Preface ................................................................... ix

Chapter 1.  Introduction

The role of models in atomic physics ................................. 1
Evidence for the existence of atoms and molecules ..................... 2
Collisions ............................................................ 3
Interaction cross-section ............................................. 6
Units, constants and symbols .......................................... 8
Relevance of atomic physics to other science disciplines and technology .... 10

Chapter 2.  Photons

Electromagnetic radiation and its properties .......................... 13
The electromagnetic spectrum .......................................... 16
The photoelectric effect .............................................. 18
Failure of classical theory ........................................... 19
Planck's theory of thermal radiation .................................. 20
Einstein's equation for the photoelectric effect ...................... 21
X-Ray radiation ....................................................... 24
Compton scattering .................................................... 25
Relativistic mechanics ................................................ 26
Conservation laws ..................................................... 28
Derivation of the Compton shift ....................................... 29
Comparison of the photoelectric effect and Compton scattering ......... 32
Electrons in metals ................................................... 34
Pair production ....................................................... 34
Passage of electromagnetic radiation through matter ................... 36
Photons and vision .................................................... 37
Dual nature of electromagnetic radiation .............................. 39

Chapter 3.  Electrons and Ions

Electrolysis .......................................................... 43
Cathode rays .......................................................... 44
Motion of charged particles in electric and magnetic fields ........... 44
Measurement of the specific charge $e/m$ for electrons ................ 50
The electronic charge $e$ ............................................. 51
The variation of electron mass with velocity .......................... 55

Positive rays ........................................................... 55
Measurement of $q/m$ for positive ions .............................. 55
Isotopes ............................................................... 57
Isotopic and atomic weights ........................................... 57
Wave–particle duality ................................................. 58
The uncertainty principle ............................................. 59
The principle of complementarity ..................................... 62

## Chapter 4.  *Models of the Atom*

Electric charges in atoms ............................................. 67
The preliminary experiments .......................................... 67
Some simple models ................................................... 69
Rutherford's model ................................................... 69
The nucleus ........................................................... 71
Limitations of Rutherford's model .................................... 74
The basic features of atomic spectra .................................. 75
Bohr's quantum postulates ............................................ 78
The planetary model .................................................. 79
The correspondence principle ......................................... 82

## Chapter 5.  *Experimental Evidence for Quantization*

Energy levels ......................................................... 89
Resonance, excitation and ionization potentials ....................... 90
The measurement of critical potentials ................................ 92
Controlled excitation of spectra ...................................... 94
Photoionization ...................................................... 95
Absorption spectra ................................................... 95
Spectra and energy levels of the alkali atoms ......................... 96
Electron spin ......................................................... 99
Angular momentum and magnetic moments ........................... 100
Stern–Gerlach experiment ............................................ 102
Zeeman effect ........................................................ 103
Further remarks about quantum numbers ............................. 104
Spin–orbit interaction ................................................ 106
Many-electron atoms ................................................. 106
X-Ray spectra ........................................................ 106
Moseley's Law ....................................................... 107
Deficiencies in the Bohr theory ...................................... 109
The exclusion principle ............................................... 110
Electron shells and the periodic table ................................ 112

## Chapter 6.  *Development of the Quantum Theory*

Criticism of the basis of classical mechanics ......................... 119
Implications of wave–particle duality ................................ 120
Mathematical representation of wave motions ......................... 120
The wave equation and its solution ................................... 123

Schrödinger's wave equation ....................................... 124
Wave packets ...................................................... 126
Interpretation of the wavefunction ................................ 128
Solution of Schrödinger's equation for some simple cases .......... 129
Further development of the quantum theory ......................... 139
Formation of chemical bonds and molecules ........................ 139

## *Chapter 7.  Physics of the Nucleus*

Constituents of nuclei ............................................ 143
Nuclear sizes and shapes .......................................... 145
Nuclear scattering and reactions .................................. 150
Nuclear energy levels ............................................. 155
Nuclear spin and magnetic moments ................................ 157
Nuclear magnetic resonance ........................................ 158
Energy balance in nuclear reactions ............................... 159
Nuclear binding energies .......................................... 160
Radioactive decay ................................................. 161
Derivation of radioactive decay laws .............................. 163
Radioactive chains ................................................ 164
Units of radioactivity ............................................ 165
Alpha decay ....................................................... 167
Beta decay ........................................................ 170
Fission ........................................................... 171
Fusion ............................................................ 173
The nuclear force ................................................. 173
Nuclear models .................................................... 175

## *Chapter 8.  Instrumentation and Applications*

Particle accelerators ............................................. 183
Synchrotron radiation ............................................. 187
Mass spectrometry and mass separation ............................ 190
Electron lenses ................................................... 194
X-Ray and electron diffraction .................................... 195
Electron microscope ............................................... 198
X-Ray radiography and tomography .................................. 201
Emission tomography ............................................... 204
Stimulated emission of radiation .................................. 204
The helium–neon laser ............................................. 205

*Index* ........................................................... 211

# Preface

This book is concerned with the structure of the atom, and with the information gained from atomic behaviour about the nature of matter, light and electricity. Deficiencies in classical theory are examined and some fundamental problems involved in the description of atomic and sub-atomic systems are exposed. The approach to quantization is more experimental than abstract, but there is an introduction to quantum theory with some key examples.

The emphasis throughout is on concepts. While the development of the subject does not follow the historical sequence, some discussion of early theories and models is given in order to illustrate the development of physics. Reference is made throughout to experiment for the purpose of testing and supporting theories or to introduce new and usually disconcerting information, so that the interplay of theory and experiment can be seen. Instrumentation and some practical applications are described in order to show that atomic theory really can be used to construct devices of practical and commercial importance.

The study of atomic physics cannot stand on its own, because it demands a sound understanding of many aspects of classical physics. Some necessary background material is covered in the introductory chapters and some mathematical formulae are introduced here, or where needed in the main text. However, this is not an omnibus textbook, and it is assumed that a first course on atomic physics would be preceded or accompanied by courses in classical wave theory, optics, electricity and mathematics. Mathematical complexity has been avoided as much as possible. Problems are provided, both of a formal analytical and of a numerical kind, and in some cases a worked solution is given with the problem. These problems are an integral part of the book and it is unlikely that a student who has not attempted the problems can be confident that he or she has fully understood the content.

The book has developed from a course given to first year undergraduates in physics, and should serve as a textbook for introductory courses for other physical sciences and engineering science. It may also serve as background material for a wider variety of courses. As a result of changes in the syllabi for A-level courses which place increasing emphasis on atomic and nuclear physics, radioactivity and medical physics, it is hoped that the book will prove useful to sixth-form teachers.

I am greatly indebted to Vivien E. Dolley who typed the manuscript and Sheila J. Rudman who prepared the graphics. Several colleagues and organizations provided new illustrations for this book; they are acknowledged in the text.

# 1 | Introduction

## THE ROLE OF MODELS IN ATOMIC PHYSICS

Most of the difficulties which we experience in the study of atomic physics arise because it is hard to imagine and comprehend a system so far removed from our normal everyday experience and scale. We must begin our study using the physics that we already know, i.e. using the laws which describe the behaviour of ordinary objects under normal conditions, but we cannot be sure without direct experimental evidence that these laws can be extrapolated to describe a system of atomic dimensions. In order to overcome these difficulties we first make *pictures* of the unfamiliar system in terms of the more familiar — we say the atom is like a billiard ball or like a planetary system — so that we are able to visualize the system and talk about it. The limitation of this procedure arises because the unfamiliar and the familiar can never match completely, otherwise there would be no unfamiliar situation to be investigated. Therefore, in the next and more sophisticated stage in the development of the subject, we consider a limited aspect of the system and use it as a basis to construct a mathematical *model* which provides a framework for the analysis and interpretation of experimental results. It is not essential that a model should provide a picture, but it is essential that it should lead to precise predictions for physical quantities which can be measured. As an example, we will consider some models of electricity:

*Fluid model, c.* 1750.   It was believed that the flow of current in a circuit could be represented by some sort of continuous fluid flowing through the conductor. This model is little more than a picture.

*Atomic model, c.* 1830.   Faraday suggested, very tentatively, that the results of experiments on electrolysis could be understood on the basis of a granular or atomic model. The modern development of this model provides the basis for the description of electrical phenomena in terms of the behaviour of individual charges.

1

*Ether strain theory*, c. 1850–c. 1900. The ether was believed to be a substance which filled all space and was very rigid, yet permitted frictionless motion through it. All electrical phenomena were attributed to some sort of strain in the ether. This is an example of a model which was thought at one time to be a genuine representation of reality but which was subsequently abandoned. It is worth noting, however, that Maxwell used the ether model to derive his famous equations which still provide a description of the large-scale phenomena of electricity and magnetism, such as the propagation of light and radio waves and the action of an electric motor.

At various times these models existed side by side. Each gave a wholly or partially correct description of the large-scale or macroscopic phenomena while giving totally different descriptions of the small-scale or microscopic phenomena. Thus, in order to distinguish between the models we have to look for a situation in which the models predict totally different results, and for this reason there are certain crucial experiments which have a very large influence on the subject. The first experiments on the properties of the electron by Millikan and J. J. Thomson (Chapter 3) and scattering experiments by Rutherford (Chapter 4) and Compton (Chapter 2) are examples of such experiments, and we shall therefore examine these original investigations even though more accurate and sophisticated versions have since been devised.

## EVIDENCE FOR THE EXISTENCE OF ATOMS AND MOLECULES

Evidence that matter is subdivided into atoms and molecules comes from both chemistry and physics. The introduction of the concept of atoms is due to Dalton who made the following postulates (1803), from which the known laws of chemical combination can be deduced:

(*i*)   The chemical elements consist of discrete particles, called *atoms*, which cannot be subdivided by any known chemical process and which preserve their identity in chemical changes.

(*ii*)   All atoms of the same *element* are identical in all respects, and different elements have atoms differing in weight.

(*iii*)   Chemical compounds are formed by the union of different elements in simple numerical proportions.

Avogadro (1811) made the important distinction between atoms and aggregates of atoms, and gave the latter the name *molecules*. He later postulated that equal volumes of gases under the same physical conditions contain the same number of molecules. The first physical evidence for the existence of these molecules was recorded in 1827 when the constant, random motion of fine particles in suspension, known as the *Brownian motion*, was observed through a microscope and interpreted as due to collisions between molecules in the liquid. The concept of molecular motion is used in the *kinetic theory of gases*, which was developed between 1740 and 1900, and is based on the following assumptions:

(*i*)   A gas consists of a very large number of molecules in rapid and random motion. The molecules are small compared with the average distances between them.

(*ii*)   The molecules are non-interacting, i.e. they undergo perfectly elastic collisions.

(*iii*)   The temperature of a gas is proportional to the average kinetic energy of the molecules.

With these assumptions it is possible to predict the gas laws, the specific heats of gases and other properties with a considerable degree of success.

## COLLISIONS

Many of the properties of an atomic system can be investigated by studying the way in which various projectiles are scattered by the system. Now if both the projectile and the target are solid impenetrable objects, like the ubiquitous billiard balls of classical physics, a collision occurs when the distance between the two centres of mass is equal to the sum of the radii (see Fig. 1.1a). Otherwise the projectile completely misses the target and continues undisturbed from its initial trajectory while the target remains at rest (Fig. 1.1b). When collisions occur, the angle of scattering and recoil and the final velocities can be calculated using the laws of conservation of energy and linear momentum. However, if the projectile and target carry charges of the same sign there is a repulsive force between them which is effective at distances much greater than the sum of the radii and may cause the projectile to be scattered (i.e. deflected from its initial direction), even though the distance of closest approach is greater than the sum of the radii (Fig. 1.1c). In this case, we say that the projectile has been scattered due to its electro-static interaction with the target. This is *action at a distance*. The same argument applies to all attractive or repulsive forces, whether magnetic, gravitational or nuclear in origin.

In *elastic scattering* there is no change in the internal energy of the target so that the energy changes occur in the kinetic energy, and the conservation

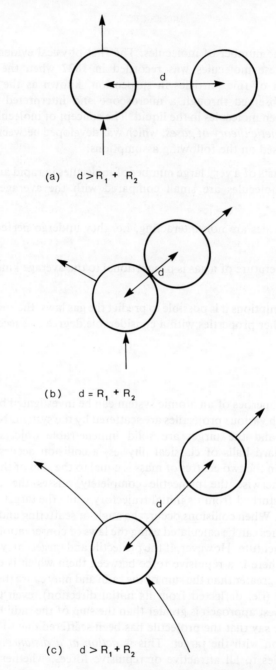

(a)  $d > R_1 + R_2$

(b)  $d = R_1 + R_2$

(c)  $d > R_1 + R_2$

*Figure 1.1.* Diagrams (a) and (b) represent billiard ball scattering in which the projectile (a) misses the target and continues undisturbed or (b) collides with the target and is deflected. Diagram (c) represents the scattering due to the action of force at a distance.

laws can be applied exactly as in billiard ball collisions. However, in *inelastic scattering*, energy is transferred to the target in a manner which can be understood only when we consider the internal structure of the target, although the total energy is still conserved. It is often convenient to refer to a scattering process between a projectile and a molecular, atomic or subatomic system as a collision, but it should be remembered that in general we are not picturing the process as a billiard ball collision between hard-edged objects.

The momentum vectors of particles involved in a collision may be specified in any convenient frame of reference. One obvious choice is the *laboratory frame* in which the target particle is at rest. We may transform to any other frame of reference which is moving with a constant velocity $U$ relative to the laboratory frame. In the *centre of mass frame* the centre of mass of the two particles is at rest, i.e. the momenta are equal and opposite.

Figure 1.2 shows the momentum vectors in the laboratory frame and the centre of mass frame for a collision in which a projectile of mass $m_1$ strikes a target of mass $m_2$ and as a result two particles of mass $m_3$ and $m_4$ emerge.

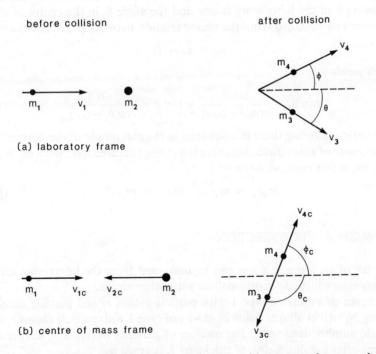

*Figure 1.2.* Collision of a projectile of mass $m_1$, with a target of mass $m_2$ leading to two particles of mass $m_3$ and $m_4$ represented in (a) the laboratory frame and (b) the centre of mass frame.

Using non-relativistic kinematics, the relation between the initial velocities in the two frames is given by

$$v_{1c} = v_1 - U, \quad v_{2c} = U$$

and since the momenta are equal in the centre of mass frame we have

$$m_1(v_1 - U) = m_2 U$$

$$\therefore U = \frac{m_1}{m_1 + m_2} v_1, \quad v_{1c} = \frac{m_2}{m_1 + m_2} v_1. \tag{1.1}$$

The energy available in the centre of mass frame is given by

$$E_c = \tfrac{1}{2} m_1 v_{1c}^2 + \tfrac{1}{2} m_c v_{2c}^2 = \frac{1}{2} \frac{m_1 m_2}{m_1 + m_2} v_1^2.$$

Thus we can write

$$E_c = \tfrac{1}{2} \mu v_1^2, \quad \mu = \frac{m_1 m_2}{m_1 + m_2} \tag{1.2}$$

where $\mu$ is called the *reduced mass*. The relation between the angle of scattering $\theta$ in the laboratory frame and the angle $\theta_c$ in the centre of mass frame can be obtained from the vector relation between the final velocities,

$$v_{3c} = v_3 - U$$

which yields

$$\tan \theta_c = \frac{\sin \theta}{\cos \theta - U/v_3}, \quad \tan \theta = \frac{\sin \theta_c}{\cos \theta_c + U/v_{3c}} \tag{1.3}$$

For elastic scattering there is no change in the magnitude of the momentum in the centre of mass frame but there is a change in direction. Since $m_1 = m_3$, $m_2 = m_4$ in this case, we have

$$m_1 v_{1c} = m_1 v_{3c}, \quad m_2 v_{2c} = m_2 v_{4c}. \tag{1.4}$$


## INTERACTION CROSS-SECTION

The size of an atom in a gas can be measured from the interaction cross-section with which the atoms collide with other atoms.

A beam of atoms of type 1 with particle radius $r_1$ and particle number density $N_1$ strikes a layer made of atoms of type 2 with particle radius $r_2$ and particle number density $N_2$. The number of particles of type 1 which are still present after passing a slab of thickness $L$ is given by

$$N = N_1 e^{-N_2 \sigma L} \tag{1.5}$$

and number of atoms which are deflected is

$$N_{scatt} = N_1(1 - e^{-N_2\sigma L}).$$  (1.6)

Here $\sigma$ is the interaction cross-section given by

$$\sigma = \pi(r_1 + r_2)^2$$  (1.7)

and corresponds to the collision illustrated in Fig. 1.1b.

The absorption of light by atoms or molecules is given by a similar formula, known as the Lambert–Beer law, i.e.

$$I(x) = I_0 e^{-N\sigma x}$$  (1.8)

where $I_0$ is the incident intensity. This is illustrated in Fig. 1.3. The absorption of X-rays or neutrons by matter obeys the same law.

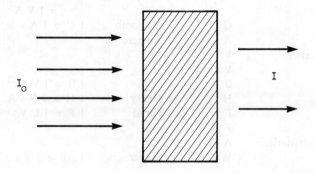

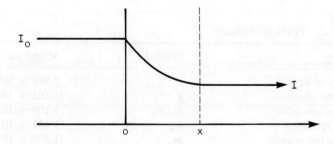

*Figure 1.3.* · Attenuation of a beam as it passes through an absorbing medium.

## UNITS, CONSTANTS AND SYMBOLS

We normally use SI uits. A list of units and symbols for some important
quantities is given in Table 1.1. The magnitude of some important constants
is given in Table 1.2.

*Table 1.1.*   SI units for important physical quantities

| Quantity | Symbol | Name | Remarks |
|---|---|---|---|
| Length | m | Metre | |
| Mass | kg | Kilogram | |
| Time | s | Second | |
| Frequency | Hz | Hertz | $1\ \text{Hz} = 1\ \text{cycle s}^{-1}$ |
| Force | N | Newton | $1\ \text{N} = 1\ \text{kg m s}^{-1}$ |
| Energy | J | Joule | $1\ \text{J} = 1\ \text{kg m}^2\ \text{s}^{-2}$ |
| Power | W | Watt | $1\ \text{W} = 1\ \text{J s}^{-1}$ <br> $= 1\ \text{V A}$ |
| Charge | C | Coulomb | $1\ \text{C} = 1\ \text{A s}$ |
| Current | A | Ampere | |
| Potential difference | V | Volt | |
| Electric field | $\text{V m}^{-1}$ | | |
| Resistance | Ω | Ohm | $1\ \Omega = 1\ \text{V A}^{-1}$ |
| Inductance | H | Henry | $1\ \text{H} = 1\ \text{V s A}^{-1}$ |
| Capacitance | F | Farad | $1\ \text{F} = 1\ \text{C V}^{-1}$ |
| Permittivity | $\text{F m}^{-1}$ | | |
| Magnetic field strength | $\text{A m}^{-1}$ | | |
| Magnetic flux | Wb | Weber | $1\ \text{wb} = 1\ \text{V s}$ |
| Magnetic induction ⎫<br>Magnetic flux density ⎭ | T | Tesla | $1\ \text{T} = 1\ \text{wb m}^{-2}$ |
| Permeability | $\text{H m}^{-1}$ | | |

*Table 1.2.*   Physical constants

| Quantity | Symbol | Magnitude |
|---|---|---|
| Velocity of light | $c$ | $2.998 \times 10^8\ \text{m s}^{-1}$ |
| Electronic charge | $e$ | $1.602 \times 10^{-19}\ \text{C}$ |
| Electron rest mass | $m_e, m_0$ | $9.109 \times 10^{-31}\ \text{kg}$ |
| Proton rest mass | $m_p$ | $1.673 \times 10^{-27}\ \text{kg}$ |
| Neutron rest mass | $m_n$ | $1.675 \times 10^{-27}\ \text{kg}$ |
| Planck's constant | $h$ | $6.626 \times 10^{-34}\ \text{J s}$ |

In atomic physics it is convenient to introduce some useful additional units. The most important of these is a unit of energy called the *electron-volt* (eV) which is defined as the energy of a singly charged particle with charge equal in magnitude to that of an electron which has been accelerated through a potential difference of one volt. In terms of the SI unit of energy, the electron-volt has the magnitude

$$1 \text{ eV} = 1.60 \times 10^{-19} \text{ J}.$$

It also convenient to use a smaller unit of length. This can be the ångström unit or the nanometre, which are defined as

$$1 \text{ Å} = 10^{-10} \text{ m} = 0.1 \text{ nm},$$

although the nanometre is now preferred.

It is often convenient to limit the powers of 10 written explicitly and to use instead multiples or sub-multiples. The appropriate prefixes and symbols for some of these multiples are given in Table 1.3. The examples most commonly occurring are

$$\text{keV} = 10^3 \text{ eV} \qquad \text{MeV} = 10^6 \text{ eV}$$

$$\text{fm} = 10^{-15} \text{ m} \qquad \text{nm} = 10^{-9} \text{ m}.$$

*Table 1.3.* Multiples and sub-multiples of units

| Multiplication factor | Prefix | Symbol |
| --- | --- | --- |
| $10^{12}$ | Tera | T |
| $10^9$ | Giga | G |
| $10^6$ | Mega | M |
| $10^3$ | Kilo | k |
| $10^{-3}$ | Milli | m |
| $10^{-6}$ | Micro | $\mu$ |
| $10^{-9}$ | Nano | n |
| $10^{-12}$ | Pico | p |
| $10^{-15}$ | Femto | f |

In mathematical formulae it is standard practice to use Greek letters to represent important physical quantities. The Greek alphabet is given in Table 1.4.

*Table 1.4.* The Greek alphabet

| A | $\alpha$ | alpha | N | $\nu$ | nu |
|---|---|---|---|---|---|
| B | $\beta$ | beta | Ξ | $\xi$ | xi |
| Γ | $\gamma$ | gamma | O | $o$ | omicron |
| Δ | $\delta$ | delta | Π | $\pi$ | pi |
| E | $\epsilon$ | epsilon | P | $\rho$ | rho |
| Z | $\zeta$ | zeta | Σ | $\sigma$ | sigma |
| H | $\eta$ | eta | T | $\tau$ | tau |
| Θ | $\theta$ | theta | Υ | $\upsilon$ | upsilon |
| I | $\iota$ | iota | Φ | $\phi$ | phi |
| K | $\kappa$ | kappa | X | $\chi$ | chi |
| Λ | $\lambda$ | lambda | Ψ | $\psi$ | psi |
| M | $\mu$ | mu | Ω | $\omega$ | omega |

## RELEVANCE OF ATOMIC PHYSICS TO OTHER SCIENCE DISCIPLINES AND TECHNOLOGY

Atomic physics is the basis for molecular physics and solid state physics, and also part of chemistry. In addition, atomic physics is a basic science for other science disciplines and for technology and applications. Some examples are shown in Fig. 1.4.

## PROBLEMS

An asterisk indicates that a problem is more difficult or more sophisticated than the general level.

**1.1.** The kinetic theory of gases can be regarded as a model of the behaviour of gases. Consider the assumptions of the kinetic theory, and state in what respect they are likely to depart from the properties of a real gas.

**1.2.** A projectile of mass $m$ and velocity $v$ strikes a target, and the magnitude and direction of the velocities of projectile and target after elastic scattering are observed. Assuming that non-relativistic kinematics are valid, show that:

(*a*) if the projectile is at rest after the collision, the mass of the projectile and the mass of the target are equal;

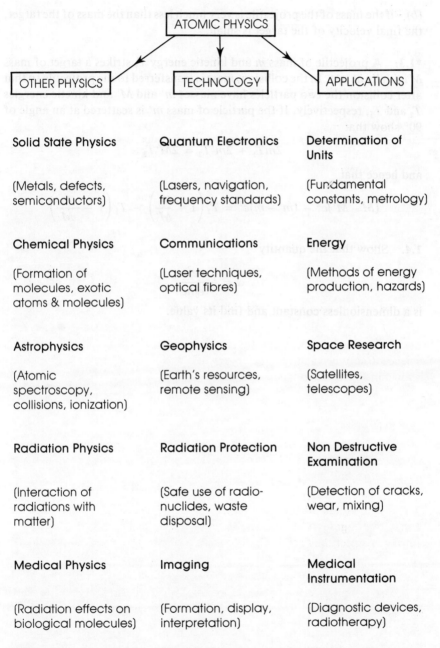

*Figure 1.4.* The relevance of atomic physics to other disciplines in science and technology.

(b)   if the mass of the projectile is very much less than the mass of the target, the final velocity of the target is small.

*1.3.   A projectile of mass $m$ and kinetic energy $T$, strikes a target of mass $M$ initially at rest. In the collision, mass is transferred to the projectile so that after collision the two particles have masses $m'$ and $M'$ and kinetic energies $T_2$ and $T_3$, respectively. If the particle of mass $m'$ is scattered at an angle of $90°$ show that

$$2mT_1 + 2m'T_2 = 2M'T_3$$

and hence that

$$(M - M')c^2 - (m' - m)c^2 = T_2\left(1 + \frac{m'}{M'}\right) - T_1\left(1 - \frac{m_1}{M}\right)$$

1.4.   Show that the quantity

$$\alpha = \tfrac{1}{2}\,\frac{e^2}{\epsilon_0 hc}$$

is a dimensionless constant and find its value.

# 2 | Photons

One of the ways in which energy can conveniently be supplied to an atom is by means of light or other electromagnetic radiation, and for this reason a study of the mechanism of absorption and emission of radiation by atoms reveals some crucial information about the structure of atoms. In this section, we examine selected aspects of the interaction between electromagnetic radiation and atoms which yield information about the nature of radiation itself. In so doing, we will introduce the concept of the *photon* as the particle of *electromagnetic radiation*.

## ELECTROMAGNETIC RADIATION AND ITS PROPERTIES

Radio waves, light and X-rays are all forms of a type of wave motion known as electromagnetic radiation. The disturbance in the medium consists of an oscillating electric field and an oscillating magnetic field, both transverse to the direction of propagation.

The classical theory of electromagnetic radiation, due to Maxwell, associates the radiation with an accelerated charge in such a way that a charge oscillating with a given frequency emits radiation of the same frequency.

All electromagnetic radiation has the same velocity in vacuo; this is the velocity of light $c$. In all other media the wavelength and velocity are changed by an amount determined by the refractive index, which is the ratio of the velocity in vacuo to the velocity in the medium. The observed variation of refractive index with wavelength is evidence of dispersion, and makes it possible to use a glass prism to disperse or spread out the components of a beam of light.

Many of the properties of electromagnetic radiation, and particularly of light, can be described in terms of *rays*. We draw rays in the direction of propagation, as shown in Fig. 2.1, and define the *wavefront* to be the surface

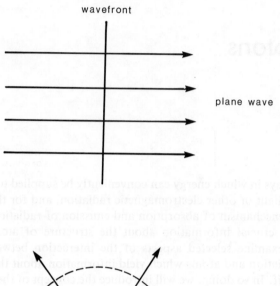

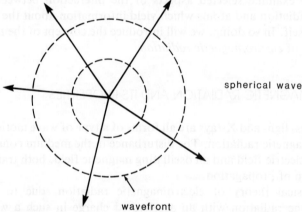

*Figure 2.1.*   Ray diagrams for a plane wave and a spherical wave, illustrating the definition of a wavefront.

perpendicular to these rays. The energy associated with the wave motion is assumed to be distributed uniformly over the wavefront, and the *intensity* of the wave is defined to be the energy flow per unit area of the wavefront per unit time, i.e. the energy flux per unit area. According to the property of *rectilinear propagation*, the rays are straight lines, and we expect that if part of the wavefront is cut off by an obstacle the rest of the wavefront will be undisturbed so that a sharp geometrical shadow of the obstacle is observed. The prediction given by ray optics for the intensity of a beam of light

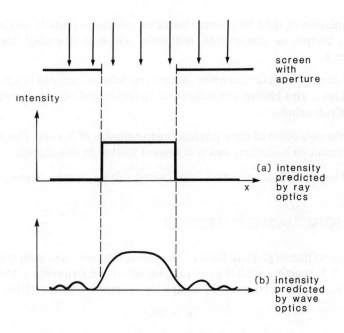

*Figure 2.2.* The prediction of the intensity falling on a screen some distance from an aperture according to (a) ray optics and (b) wave optics.

observed on a distant screen after it has passed through an aperture is shown in Fig. 2.2a. However, it is found that if the size of the aperture is of the same order of magnitude as the wavelength of the light, a more complicated pattern of bright and dark regions is observed, as shown in Fig. 2.2b. This is an example of the phenomenon of *diffraction* by a single slit, which can be interpreted using wave optics in terms of the phase difference between the disturbance due to different parts of the secondary wave propagated from the aperture. Using the *principle of superposition* it can be shown that the bright regions of the diffraction pattern correspond to disturbances which differ in phase by $2\pi$, which implies a path difference equal to an integral number of wavelengths, while the dark regions correspond to phase differences of $\pi$ or a path difference of odd half-integer numbers of wavelengths. Similarly, the wave theory can be used to explain the *interference* patterns observed in two-slit and similar experiments.

There are some other means of generating electromagnetic waves besides the oscillating charge. These are:

(*i*)   Emission of light by charged particles moving in particle accelerators. This is known as *synchrotron radiation*, which is discussed further in Chapter 8.

(*ii*)   Emission of radiation when charged particles are trapped in a magnetic field. This is also known as synchrotron radiation and occurs, for example, in the Crab nebula.

(*iii*)   Slowing down of electrons leading to emission of X-rays. This is called *bremsstrahlung radiation*, and is discussed further in this chapter.

(*iv*)   Thermal radiation, which is discussed further in this chapter.

## THE ELECTROMAGNETIC SPECTRUM

The various names given to parts of the electromagnetic spectrum are shown in Fig. 2.3, together with the relevant values of the *frequency ν* and *wavelength* λ, which are connected through the spectrum by the relation

$$c = \nu\lambda, \tag{2.1}$$

where *c* is the velocity of light. The components of the electromagnetic spectrum have the following characteristics:

(*i*)   *Radio frequency waves.* These have wavelengths from a few km to about 0.3 m. The frequency range is a few Hz to $10^9$ Hz. They are used in television and broadcasting systems and are generated by electronic devices.

(*ii*)   *Microwaves.* These have wavelengths from 0.3 m down to $10^{-3}$ m, and the frequency range is from $10^9$ Hz to $3 \times 10^{11}$ Hz. They are used in radar and other communication systems, and are generated by electronic devices and certain molecular transitions.

(*iii*)   *Infrared radiation.* This covers the wavelengths from $10^{-3}$ m down to $7.8 \times 10^{-7}$ m, and the frequency range is $3 \times 10^{11}$ Hz to $4 \times 10^{14}$ Hz. These waves are generated in electronic or molecular transitions and as thermal radiation from hot bodies.

(*iv*)   *Light.* The visible spectrum covers a narrow band of wavelengths from $7.8 \times 10^{-7}$ m to $3.8 \times 10^{-7}$ m to which the retina of the human eye is sensitive. The frequency range is $4 \times 10^{14}$ Hz to $8 \times 10^{14}$ Hz. The radiation is generated by transitions in atoms and molecules. Light of a definite wavelength produces a characteristic sensation in the eye which we call *colour*. For this reason an electromagnetic wave of well-defined wavelength is called *monochromatic*.

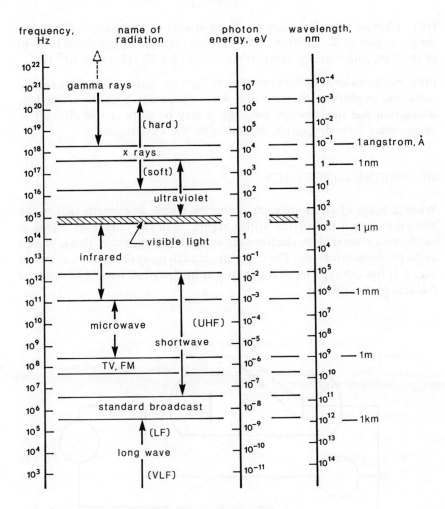

*Figure 2.3.* The spectrum of electromagnetic radiation.

(*v*)  *Ultraviolet radiation.* This covers wavelengths from $3.8 \times 10^{-7}$ m down to about $6 \times 10^{-10}$ m, or a frequency range of $8 \times 10^{14}$ Hz to $5 \times 10^{17}$ Hz. This radiation is generated by transitions in atoms and molecules, and is responsible for many of the chemical effects of electromagnetic radiation.

(*vi*)  *X-Rays.* This radiation extends from wavelengths of $10^{-9}$ m down to $6 \times 10^{-12}$ m, or frequencies of $3 \times 10^{17}$ Hz to $5 \times 10^{19}$ Hz, and is generated by the highest-energy transitions in atoms or by the deceleration of high-energy electrons.

(*vii*) *Gamma rays*. These are produced in nuclear transitions and overlap the X-ray part of the spectrum. The wavelengths extend from about $10^{-10}$ m to $10^{-14}$ m, with corresponding frequencies of $3 \times 10^{18}$ Hz to $3 \times 10^{22}$ Hz.

From our knowledge of the properties of light we expect all electromagnetic radiation to display characteristic wave properties such as refraction, diffraction and interference, although it may be more or less difficult to demonstrate these properties, depending on the wavelength.

## THE PHOTOELECTRIC EFFECT

When a beam of electromagnetic radiation whose wavelength falls in the short-wavelength end of the visible region or in the ultraviolet region is incident on a metal plate, electrons are emitted from the plate. This is known as the photoelectric effect. The apparatus used to study this effect is shown in Fig. 2.4, but the experimental details will be described later. The essential features are as follows:

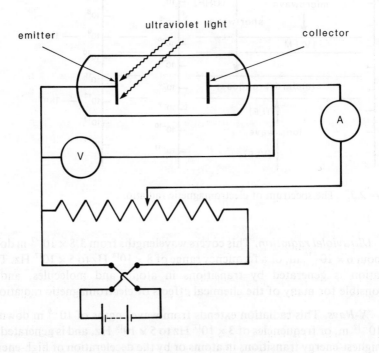

*Figure 2.4.*   An apparatus for the study of photoelectric effect.

(*i*) The number of electrons emitted per unit time is proportional to the intensity of the radiation.

(*ii*) The velocities of the emitted electrons have values between zero and a definite maximum. The proportion of the electrons having a particular velocity is independent of the intensity of the radiation.

(*iii*) The maximum velocity observed depends on the frequency of the radiation but not on the intensity.

(*iv*) For each type of emitting material there is a threshold frequency below which no electrons are emitted however great the intensity of the radiation.

(*v*) Electrons are emitted almost instantaneously even when the intensity is very low. The measured time lag between the time when the radiation first falls on the emitting material and the emission of the first electrons is less than $10^{-8}$ s.

These observations indicate that it is the frequency and not the intensity of the radiation which determines the velocities and hence the energies of the emitted electrons.

## FAILURE OF CLASSICAL THEORY

According to classical wave theory the energy carried by a wave motion is distributed uniformly and continuously over a wave front and is measured by the intensity. Thus, when radiation falls on a metal plate the energy of the wave should be transferred uniformly to the electrons in the surface and the amount of energy taken up by the electrons should be proportional to the intensity of the wave. The process of transfer of energy to the electrons should be independent of the frequency. If the intensity is low, a longer time should elapse before the electrons acquire sufficient energy to break free from the metal plate. It is evident that the classical theory completely fails to account for the observations (ii)–(iv) listed above.

The magnitude of the discrepancy between the theoretical predictions and the experimental observations can be illustrated by considering the effect of very low intensity radiation. The emission of electrons from sodium can be detected as a result of irradiation by the light from a candle placed 10 m away from the metal and the time lag for this emission is still of the order of $10^{-8}$ s. In contrast, a calculation of the time taken for electrons to absorb sufficient energy to escape from the metal when subject to irradiation at such a low intensity predicts a time lag of 1 s. This discrepancy can only be resolved by the assumption of an entirely new mechanism for the absorption of energy in the photoelectric process. It must be remembered, however, that the wave theory does give a very satisfactory description of many properties of electromagnetic radiation, such as interference phenomena, transmission of radio waves and so on.

## PLANCK'S THEORY OF THERMAL RADIATION

At about the time when the first studies of the photoelectric effect were being carried out (1887–1905) there was also considerable interest in the process of emission of thermal radiation by hot bodies. This thermal radiation is electromagnetic radiation in the infrared region. Figure 2.5 shows the spectral distribution $I(\lambda)$ of this radiation, i.e. the energy emitted per unit area per unit time by a body at temperature $T$ in the form of radiation with a wavelength between $\lambda$ and $\lambda + d\lambda$. The classical theory for this process was formulated by Rayleigh and Jeans, and completely fails to reproduce the experimental results except in the limit of long wavelengths.

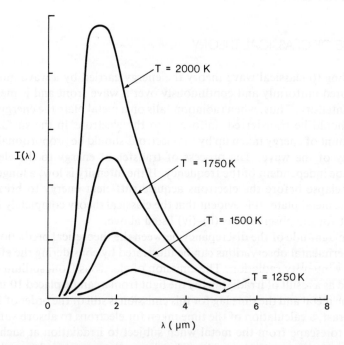

*Figure 2.5.*   The spectral distribution $I(\lambda)$ of thermal radiation emitted by a hot body at various values of the temperature $T$.

The disagreement between theory and experiment was resolved by Planck (1901) who postulated that the *emission and absorption of radiation is not a continuous process*, but instead the energy changes occur as integral multiples of a basic unit called the *quantum* of energy. The energy of the quantum is given by

$$E = h\nu, \tag{2.2}$$

where $\nu$ is the frequency and $h$ is a universal constant known as *Planck's constant*. Using this postulate together with the standard techniques of thermodynamics and statistical mechanics, Planck succeeded in deriving a formula which reproduced the experimental results and so obtained the first estimate of the constant $h$. The accepted magnitude of this constant, which has dimensions of action (energy × time), is now $6.625 \times 10^{-34}$ J s. In the long-wavelength limit, the frequency decreases so that the energy changes appear more nearly continuous and we may reasonably expect that the predictions of Planck's quantum theory should approach the predictions of the classical formula. This explains why the Rayleigh–Jeans theory has the correct behaviour in the long-wavelength limit.

## EINSTEIN'S EQUATION FOR THE PHOTOELECTRIC EFFECT

Einstein (1905) extended Planck's quantum theory of thermal radiation and applied it to the photoelectric effect. He made the following assumptions:

(*i*) Electromagnetic radiation of frequency $\nu$ consists of quanta of energy called *photons* which have energy $E = h\nu$ and travel with the velocity of light.

(*ii*) In the photoelectric effect one photon is completely absorbed by one electron which thereby gains the quantum of energy and may be emitted from the metal.

With these assumptions the qualitative observations in the photoelectric effect can easily be predicted. The electrons are initially bound in the metal, but if the quantum of energy $h\nu$ absorbed by an electron exceeds the magnitude of the binding energy, it is released from the metal and takes up the balance of the energy in the form of kinetic energy. Thus the maximum kinetic energy observed is given by

$$\tfrac{1}{2}mv_{max}^2 = h\nu - W \tag{2.3}$$

where the *work function* $W$ is the *minimum* energy required to release an electron from the surface of the metal. This equation clearly predicts that the maximum kinetic energy is directly proportional to the frequency, and that there must be a threshold frequency $\nu_0$ such that $h\nu_0 = W$ when the photon energy is just sufficient to overcome the binding of the least-bound electrons. This threshold frequency will depend on the work function for a particular material, but if the frequency of the radiation is below this value no emission can occur. An increase in the intensity of the radiation is equivalent to an increase in the number of photons falling on the emitting surface, and if the frequency is above the threshold value, i.e. $\nu > \nu_0$, this will increase the number of electrons emitted, but when $\nu < \nu_0$ an increase in the intensity cannot have any effect. However much the intensity is reduced, whenever a photon with $\nu > \nu_0$ arrives at the surface there is a chance that photoelectric emission can occur, and there is no time delay.

Equation (2.3) gives the maximum kinetic energy of the emitted electrons. Electrons which are emitted with lower energies may have suffered collisions inside the metal before leaving the surface, or may initially have had a binding energy slightly larger than $W$. The first quantitative experimental verification of Einstein's equation was carried out by Millikan in 1916. The equation can also be verified using the apparatus shown in Fig. 2.4. First, by making the potential $V$ of the collecting plate positive with respect to the emitting plate, it is possible to draw a steady current of electrons across the gap between the plates and so to investigate the effect of varying the intensity of the radiation for a fixed frequency. If the collecting plate is at a negative potential with respect to the emitter the photoelectrons are now repelled by the collector, and for a sufficiently large retarding potential even the most energetic electrons fail to reach the collector so that no current flows. The behaviour of the current as a function of the potential $V$ is shown in Fig. 2.6a. The potential which is just sufficient to stop the most energetic electrons is known as the *stopping potential* $V_s$ and must be connected with the maximum kinetic energy through the relation

$$V_s e = \tfrac{1}{2} m v_{max}^2.$$

Next, by measuring the stopping potential for a range of frequencies it is possible to obtain the linear graph shown in Fig. 2.6b. This line is found to have a gradient of $h/e$ and intercept $\nu_0$. Thus

$$V_s = \frac{h}{e}(\nu - \nu_0). \tag{2.4}$$

Finally, by repeating these measurements with different emitters it can be demonstrated that the gradient of the graph of $V_s$ against $\nu$ is unaffected but the threshold frequency depends on the nature of the emitter. Hence, setting

$$h\nu_0 = W \tag{2.5}$$

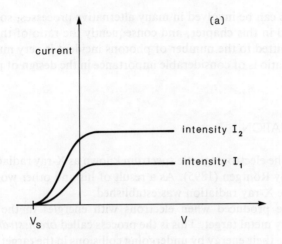

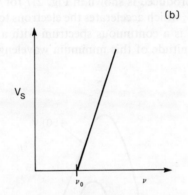

*Figure 2.6.* Results obtained from a study of the photoelectric effect. (a) The variation of the photoelectric current with the potential on the collecting plate. (b) The variation of the stopping potential with the frequency of the electromagnetic radiation.

and combining eqns (2.3)–(2.5), this group of experiments serves to verify Einstein's equation (2.2).

It should be noted that Einstein's theory of the photoelectric effect implies that the emission of one electron is associated with the absorption of one photon. It does not follow that every photon incident on the metal causes emission of an electron, even though the photon energy is above threshold.

The photons can be involved in many alternative processes, some of which are discussed in this chapter, and consequently the ratio of the number of electrons emitted to the number of photons incident is very much less than unity. This ratio is of considerable importance in the design of photoelectric devices.

## X-RAY RADIATION

The part of the electromagnetic spectrum known as X-ray radiation was first discovered by Röntgen (1895). As a result of his and other work, the wave nature of the X-ray radiation was established.

X-rays are produced when electrons with energies in the keV region strike a heavy metal target. This is the process called *bremsstrahlung*. These electrons lose their energy by undergoing collisions in the target, and at each collision some or all of the lost energy appears as radiation. The spectrum of the X-ray radiation produced is shown in Fig. 2.7 for various values of the accelerating voltage $V$ which accelerates the electrons to an energy eV, and it can be seen that this is a continuous spectrum with a sharp cut-off wavelength $\lambda_{min}$. The magnitude of this minimum wavelength can be calculated

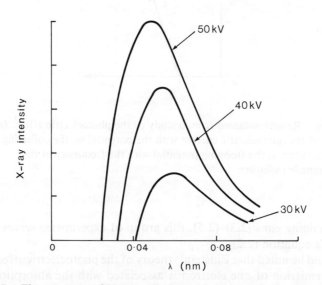

*Figure 2.7.* The spectrum of X-ray radiation emitted from an X-ray tube at various accelerating voltages.

using the quantum concept. According to this theory, each X-ray photon is produced as the result of a particular collision involving one electron, and the photons of maximum energy are produced when an electron loses the whole of its energy in a single collision. (This process may be thought of as an inverse photoelectric effect.) For an accelerating voltage $V$, the maximum photon energy is given by

$$h\nu_{max} = \tfrac{1}{2}mv^2 = eV,$$

but a maximum frequency corresponds to a minimum wavelength so that

$$\lambda_{min} = \frac{hc}{eV}. \tag{2.6}$$

The magnitude of $\lambda_{min}$ can be measured accurately, and eqn (2.6) can be used to derive a value for Planck's constant which is in agreement with the values obtained from the studies of thermal radiation and the photoelectric effect.

Events in which the electron gives up the whole of its energy are rare, and more usually an electron produces several photons of wavelength $\lambda > \lambda_{min}$ in a succession of collisions. The electrons also cause some ionization and much of the kinetic energy (sometimes up to 99%) is converted into heat. In addition to the smooth continuous spectrum displayed in Fig. 2.7, some sharp X-ray lines can appear at sufficiently high accelerating voltages; the mechanism for their production is discussed in Chapter 5.

## COMPTON SCATTERING

The scattering of monochromatic X-rays from targets composed of light elements was first studied by Compton (1923) who found that the scattered radiation consists of two lines, one of the same wavelength as the incident radiation and one of slightly longer wavelength. This phenomenon of Compton scattering can be explained by treating it as an elastic collision between the X-ray photon and a loosely bound electron.

Consider the scattering process shown in Fig. 2.8 in which the incident photon has frequency $\nu$ and wavelength $\lambda$. The scattered photon is detected at angle $\phi$ with frequency $\nu'$ and wavelength $\lambda'$. The electron is assumed to be initially at rest and to have a binding energy very much less than the photon energy. After collision the electron recoils at an angle $\theta$ to the direction of the incident beam with momentum $mv$.

We now need to digress in order to find out how to deal with an object with velocity $c$.

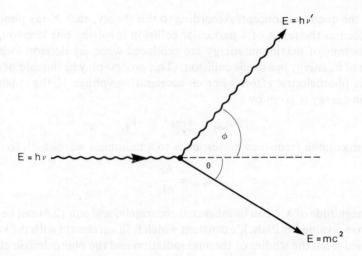

*Figure 2.8.* The kinematics of Compton scattering of a photon from a free electron.

## RELATIVISTIC MECHANICS

In classical mechanics it is necessary to specify the momenta and co-ordinates relative to some set of co-ordinate axes or *frame of reference*. According to Newtonian mechanics, if we transform from one frame of reference to another moving in the $x$-direction with velocity $u$, the co-ordinates and velocity transform according to the equations

$$x' = x - ut, \quad y' = y, \quad z' = z,$$
$$t' = t, \quad v' = v - u.$$

However, experiments show that the velocity of light measured in different frames of reference is always the same, i.e. it is always $c$. Following the first of these experiments, by Michelson and Morley, Einstein postulated that the velocity of light is independent of the motion of its source, and that the laws of electromagnetic phenomena and of mechanics are the same in all inertial frames of reference. (An inertial frame is one in which a body at rest and not under the influence of any force, will remain at rest.) Einstein then modified classical mechanics in the light of these postulates.

For our present purposes, the essential features of Einstein's theory are the following:

(*i*) The mass of a particle is a function of its velocity according to the relation

$$m = m_0/(1 - v^2/c^2)^{1/2}. \qquad (2.7)$$

This preserves the classical definition of momentum as

$$p = mv \qquad (2.8a)$$

$$= m_0 v/(1 - v^2/c^2)^{1/2}. \qquad (2.8b)$$

Here $m_0$ is called the *rest mass* and $m$ is called the *relativistic* or *moving mass*. Alternatively, one may take the definition of the momentum as fundamental and define mass through eqns (2.8a) and (2.8b); in this approach $m$ is often called the *momental mass*.

(*ii*) It follows from eqns (2.7) and (2.8) that the particle velocity can approach the value $c$ as a limiting value but can never exceed it.

(*iii*) The kinetic energy of a particle is given by

$$T = m_0 c^2/(1 - v^2/c^2)^{1/2} - m_0 c^2$$

$$= mc^2 - m_0 c^2.$$

The quantity $m_0 c^2$ is called the *rest mass energy*.

(*iv*) The total energy $E$ is defined as the sum of the kinetic energy and rest mass energy, i.e.

$$E = T + m_0 c^2 \qquad (2.9)$$

$$= mc^2. \qquad (2.10)$$

To this total energy must be added any potential energy appropriate to the system.

A useful relation between $E$ and $p$ can be obtained by squaring eqn (2.10) and substituting from eqns (2.7) and (2.8). This gives

$$E^2 = m^2 c^4 = m_0^2 c^4/(1 - v^2/c^2)$$

$$= m_0^2 c^4/(1 - p^2/m^2 c^2) = m_0^2 c^4/(1 - p^2 c^2/E^2),$$

and rearranging the last line we have

$$E^2 = p^2 c^2 + m_0^2 c^4. \qquad (2.11)$$

In order to make the connection with non-relativistic mechanics, we examine the expression (2.9) for the limiting case when $v \ll c$. We have

$$T = E - m_0c^2$$

$$= (p^2c^2 + m_0^2c^4)^{1/2} - m_0c^2 \qquad (2.12)$$

$$= m_0c^2(1 + p^2/m_0^2c^2)^{1/2} - m_0c^2,$$

and when $v^2 \ll c^2$, i.e. $p^2 \ll m_0^2c^2$, the square root can be expanded to give

$$T = m_0c^2\left\{1 + \frac{p^2}{2m_0^2c^2} + \ldots\right\} - m_0c^2$$

$$= \frac{p^2}{2m_0} + \ldots.$$

Hence, using eqns (2.7) and (2.8a) we have, for $v \ll c$,

$$m = m_0$$

$$T = \frac{p^2}{2m} = \tfrac{1}{2}mv^2. \qquad (2.13)$$

Thus, it is only in the limit of very low particle velocities that the familiar formula $T = \tfrac{1}{2}mv^2$ applies.

## CONSERVATION LAWS

In order to make quantitative statements about the energy changes, we make use of the *law of conservation of energy* which states that in an isolated system the total amount of energy remains constant although transformations from one form of energy to another may occur. In view of the relativistic relationship between energy and mass expressed in eqns (2.9) and (2.10) we must interpret mass as a form of energy in this context, so that the total energy is given by

$$W = E + V = T + V + m_0c^2. \qquad (2.14)$$

If the rest mass of the system remains unchanged we can often incorporate the rest mass energy into the potential energy and write

$$W = T + V, \qquad (2.15)$$

which is the usual form for non-relativistic problems. The difference between these two expressions essentially represents a different choice of the zero of energy, which is permissible because the definition of the potential energy always includes an arbitrary constant.

In many problems in atomic physics we are concerned with the energy with which two or more particles are bound together. As an example, we consider a system composed of a particle of small mass and another particle of infinitely heavy mass, and define the energy of this system to be zero when both particles are at rest and there is no force between them. If energy is now supplied to the system the total energy will increase and the light particle will acquire positive kinetic energy. However, if we add an attractive force between the particles this gives the system negative potential energy so that it is possible for the total energy to become negative. We interpret this negative energy as a *binding energy*, i.e. the two particles are bound together due to the attractive force and it is necessary to supply energy in order to overcome the binding and raise the system to zero or a positive energy.

In the study of collisions in which the particles are deflected or scattered from their initial path into a final path, we also make use of the *law of conservation of linear momentum* which states that when the vector sum of external forces acting on a system is zero, the total linear momentum of the system remains constant although the distribution of the total momentum among the components of the system may change.

## DERIVATION OF THE COMPTON SHIFT

Assuming that the photon can be thought of as "particle-like" with velocity $c$, it follows from eqn (2.7) that

$$m_0 = 0, \tag{2.16}$$

i.e. the rest mass of the photon must be zero. From the general relation between energy, momentum and rest mass energy, i.e. eqn (2.11), it follows that for the photon

$$E = pc \tag{2.17}$$

$$\therefore p = \frac{E}{c} = \frac{h\nu}{c} = \frac{h}{\lambda} \tag{2.18}$$

and, using this formula, Table 2.1 can be completed. It may be noted that even though the rest mass of the photon is zero, a moving or momental mass may be defined through the relation $m = E/c^2$.

The description of the Compton effect can now be obtained by applying the laws of conservation of energy and linear momentum. Conservation of energy requires that

$$h\nu + m_0 c^2 = h\nu' + mc^2. \tag{2.19}$$

*Table 2.1.*  Magnitudes of energy and momentum in Compton scattering

|                   | Electron | Photon |
|-------------------|----------|--------|
| Initial energy    | $m_0 c^2$ | $h\nu = \dfrac{hc}{\lambda}$ |
| Final energy      | $mc^2 = \dfrac{m_0 c^2}{\sqrt{1 - v^2/c^2}}$ | $h\nu' = \dfrac{hc}{\lambda'}$ |
| Initial momentum  | 0 | $\dfrac{h\nu}{c} = \dfrac{h}{\lambda}$ |
| Final momentum    | $mv = \dfrac{m_0 v}{\sqrt{1 - v^2/c^2}}$ | $\dfrac{h\nu'}{c} = \dfrac{h}{\lambda'}$ |

In order to apply the law of conservation of momentum we resolve the momenta along the direction of the incident photon and perpendicular to it. This gives

$$\frac{h\nu}{c} = \frac{h\nu'}{c} \cos \phi + mv \cos \theta \tag{2.20}$$

$$0 = \frac{h\nu'}{c} \sin \phi - mv \sin \theta. \tag{2.21}$$

By rearranging these equations to eliminate the velocity and recoil angle of the electron we obtain the solution

$$\lambda' - \lambda = \frac{h}{m_0 c} (1 - \cos \phi), \tag{2.22}$$

which gives the wavelength of the scattered photon in terms of the incident wavelength and the angle at which the scattered photon is detected. This equation was derived by Compton and verified by him experimentally.

The derivation of eqn (2.22) proceeds as follows. Equation (2.19) can be rewritten in the form

$$h\nu - h\nu' + m_0 c^2 = mc^2$$

$$\frac{hc}{\lambda} - \frac{hc}{\lambda'} + m_0 c^2 = \frac{m_0 c^2}{\sqrt{1 - v^2/c^2}},$$

and then dividing by $c$ and squaring both sides of the equation we obtain

$$\frac{h^2}{\lambda^2} + \frac{h^2}{\lambda'^2} - \frac{2h^2}{\lambda\lambda'} + 2m_0 ch \left( \frac{1}{\lambda} - \frac{1}{\lambda'} \right) + m_0^2 c^2 = \frac{m_0^2 c^2}{1 - v^2/c^2},$$

which can be rearranged to give

$$\frac{h^2}{\lambda^2} + \frac{h^2}{\lambda'^2} - \frac{2h^2}{\lambda\lambda'} + \frac{2m_0 ch}{\lambda\lambda'} (\lambda' - \lambda) = \frac{m_0^2 v^2}{1 - v^2/c^2}. \qquad (2.23)$$

Equations (2.20) and (2.21) can be rewritten in the form

$$\frac{h}{\lambda} - \frac{h}{\lambda'} \cos \phi = \frac{m_0 v}{\sqrt{1 - v^2/c^2}} \cos \theta$$

$$\frac{h}{\lambda'} \sin \phi = \frac{m_0 v}{\sqrt{1 - v^2/c^2}} \sin \theta$$

so that by squaring and adding these equations we eliminate $\theta$. This gives

$$\frac{h^2}{\lambda^2} + \frac{h^2}{\lambda'^2} - \frac{2h^2}{\lambda\lambda'} \cos \phi = \frac{m_0^2 v^2}{1 - v^2/c^2}. \qquad (2.24)$$

Now, subtracting eqn (2.24) from eqn (2.23) gives

$$- \frac{2h^2}{\lambda\lambda'} (1 - \cos \phi) + \frac{2m_0 ch}{\lambda\lambda'} (\lambda' - \lambda) = 0,$$

and hence we obtain

$$\lambda' - \lambda = \frac{h}{m_0 c} (1 - \cos \phi),$$

which is the result previously stated in eqn (2.22).

The quantity $h/m_0 c$ has dimensions of length and is known as the Compton wavelength of the electron. Its magnitude is $2.43 \times 10^{-12}$ m or 0.00243 nm. The maximum value of $(1 - \cos \phi)$ is 2 when $\phi = 180°$, and it therefore follows from eqn (2.22) that the Compton effect can most readily be detected for radiation whose wavelength is not greater than a nanometre. For example, for $\lambda = 0.5$ nm there is a maximum change in the wavelength of 1%, while for $\lambda = 0.1$ nm there is a 5% effect. We have assumed that the electron is so loosely bound to the atom that it can be regarded as essentially free. If this is not the case, and instead the electron remains tightly bound to the atom, the whole atom recoils as a result of the Compton scattering. The conservation equations are unchanged except that the rest mass $m_0$ of the electron must be replaced by the rest mass $M_0$ of the atom. For an aluminium target we have

$$M_0 \approx 27 \times \frac{m_H}{m_0} \times m_0 \approx 27 \times 1840 \times m_0,$$

where $m_H$ is the mass of the hydrogen atom, so that

$$\frac{h}{M_0 c} \approx \frac{2.43 \times 10^{-12}}{27 \times 1840} \approx 4.9 \times 10^{-17} \text{ m}$$

$$\approx 4.9 \times 10^{-8} \text{ nm.}$$

Thus, for an incident wavelength of the order of a nanometre the change in the wavelength due to scattering from tightly bound electrons is negligible for all values of $\phi$. This process gives rise to the unmodified line observed by Compton.

## COMPARISON OF THE PHOTOELECTRIC EFFECT AND COMPTON SCATTERING

In Compton scattering it is necessary that the energy of the photon is very much greater than the binding energy of the electron, i.e. $h\nu \gg W$, so that the electron behaves as if it is free. After the scattering process has taken place, a recoiling electron and a scattered photon can be observed. If we consider the possibility that the incident photon is completely absorbed and no scattered photon appears, eqns (2.23) and (2.24) become

$$\frac{h^2}{\lambda^2} + 2m_0 c \frac{h}{\lambda} = \frac{m_0 v^2}{1 - v^2/c^2}$$

and

$$\frac{h^2}{\lambda^2} = \frac{m_0 v^2}{1 - v^2/c^2},$$

and it is evident that these equations cannot simultaneously be satisfied. This means that the scattered photon must be emitted in order to conserve energy and momentum.

In the photoelectric effect the photon energy is of the same order of magnitude as the binding energy of the electron, i.e. $h\nu \approx W$. When the photon is absorbed the energy acquired serves to break this binding and both the electron and the remaining ion recoil with a given energy and momentum. The velocities involved are normally non-relativistic so that we may write the kinetic energies of the electron and the ion respectively, as

$$T_e = \frac{p_e^2}{2m_0}, \quad T_A = \frac{p_A^2}{2M_0}.$$

Since $M_0 \gg m_0$, it follows that even if $p_A \approx p_e$, $T_A \ll T_e$. Thus the ion can recoil with momentum $p_A$ such that momentum is conserved, but the cor-

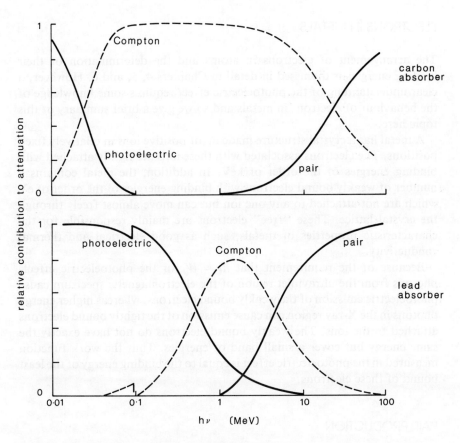

*Figure 2.9.* The relative contribution of various processes to the attenuation of a beam of electromagnetic radiation passing through matter as a function of the photon energy. (From E. Segre, *Nuclei and Particles*. Benjamin, 1964.)

responding kinetic energy is negligibly small and eqn (2.3) still effectively describes the energy balance for the process.

When a beam of electromagnetic radiation is passed through matter both the photoelectric effect and Compton scattering can occur and both lead to a loss of intensity of the beam. The relative contributions of these processes to the total attenuation of a beam is shown in Fig. 2.9 for a light and a heavy element as absorber.

## ELECTRONS IN METALS

The arrangement of electrons in atoms and the determination of their binding energies is discussed in detail in Chapters 4, 5, and 6. However, a clear understanding of the photoelectric effect requires some knowledge of the behaviour of electrons in metals and so we give a brief summary of this topic here.

A metal has a crystal structure made up of positive ions in relatively fixed positions. The electrons associated with these ions are tightly attached with binding energies of the order of keV. In addition, the metal contains a number of weakly bound electrons with binding energies of the order of eV which are not attached to any one ion but can move almost freely through the crystal lattice. These "free" electrons are mainly responsible for the characteristic properties of metals, such as good electrical and thermal conductivity.

Because of the requirement that $h\nu \approx W$ for the photoelectric effect, photons from the ultraviolet region of the electromagnetic spectrum cause photoelectric emission of the weakly bound electrons, whereas higher energy photons in the X-ray region can cause emission of the tightly bound electrons attached to the ions. The weakly bound electrons do not have exactly the same energy but cover a small band of energies. Thus the work function measured in the photoelectric effect is equal to the binding energy of the least bound of these electrons.

## PAIR PRODUCTION

A beam of high-energy photons incident on an absorber may also give rise to the process of pair production, in which one photon is converted into an electron and its *antiparticle*, the positron. If we use the symbol $\gamma$ to represent the photon, the pair production process can be written as

$$\gamma \rightarrow e^- + e^+. \tag{2.25}$$

The law of conservation of charge requires that there must be two particles of opposite charge produced, but apart from the opposite sign of its charge the positron can be regarded in this context as identical to the electron. The law of conservation of energy imposes the condition that there is a threshold energy of $2m_0c^2$ for this process, and if $h\nu > 2m_0c^2$ the balance of the energy appears as kinetic energy of the particles. Owing to the requirement that momentum should be conserved, the process cannot occur in free space but must occur near to an atom which can recoil and take up momentum. In order to demonstrate this last point, we consider that the electron and positron are produced in free space with equal momenta $p$ at equal angles $\theta$ to the direction of the incident beam. Then the equation for the conservation of momentum is

$$\frac{h\nu}{c} = 2p \cos \theta$$

or

$$h\nu = 2mc^2 \frac{v}{c} \cos \theta,$$

and the equation for energy conservation is

$$h\nu = 2m_0c^2 + 2T = 2mc^2,$$

and since $(v/c) \cos \theta < 1$, these equations cannot be satisfied simultaneously. The contribution of pair production to the attenuation of a beam is also shown in Fig. 2.9. Since $m_0c^2$ for the electron is 0.51 MeV, the threshold for pair production is 1.02 MeV, as can be seen from the figure.

A positron produced in matter will be slowed down by collisions until it eventually interacts with an electron and undergoes *pair annihilation*, to produce one or more photons. If the electron is tightly bound to an atom the whole atom can recoil to allow energy and momentum conservation, even in the case when only one photon is produced, so that this single-photon process is the exact inverse of the pair production process represented by eqn (2.25). If the electron is free, pair annihilation can still take place but the conservation laws require that more than one photon is emitted (see Question 2.8). Pair production and annihilation of other particle–antiparticle pairs have been observed, but much more energy is involved in these cases because all other fundamental particles of non-zero rest mass are much heavier than the electron. The existence of stable antiparticles has led to the speculation that other galaxies, far distant from our own, could consist of anti-matter.

In the preceding discussion of the photoelectric effect, Compton scattering, pair production and pair annihilation, our primary purpose has been to obtain simple information about the nature of electromagnetic radiation and to examine the effect of conservation laws. A detailed study of the interaction between a photon and a charged particle is well beyond the scope of this book. However, it is appropriate to note that the photoelectric effect, pair production, and pair annihilation producing a single photon, all require the presence of an atom not only to satisfy the conservation of energy and momentum but also to provide a Coulomb field in the region where the

*Table 2.2.* Energy and $Z$-dependence of photon interactions

| Process | $Z$ dependence | Important energy region | $E$ dependence |
|---------|----------------|-------------------------|----------------|
| Rayleigh scattering | $Z^{2-3}$ | < 0.5 MeV | $E^{-2}$ |
| Thomson scattering | $Z^{1-2}$ | Independent of $E$ | |
| Compton scattering | $Z$ | < 10 MeV | $E^{-1}$ |
| Photoelectric effect | $Z^{4-5}$ | < 0.5 MeV | $E^{-3}$ |
| Pair production | $Z^2$ | > 1.02 MeV | $\log E$ |

photon and electron interact. For this reason, the relative importance of these processes depends on the type of atom as well as the energy. This can be seen from the comparison of absorption processes in a light and a heavy atom, shown in Fig. 2.9.

From what we have done so far it is not obvious that a neutral atom can produce a Coulomb field. To understand this we need Rutherford's nuclear model of the atom which is introduced in Chapter 4. It will then be seen that the field strength depends on the nuclear charge $+ Ze$. The processes described above depend on different powers of $Z$, as shown in Table 2.2. For Compton scattering, the total intensity of the scattered radiation from an atom is just $Z$ times the scattering from a single electron, provided the condition $h\nu \gg W$ is satisfied.

The energy dependence of these processes is also given in Table 2.2. From this table and from Fig. 2.9, it can be seen that pair production dominates at high photon energies and high $Z$. The Compton effect dominates at medium energies and low $Z$, and the photoelectric effect dominates at low energies and high $Z$.

## PASSAGE OF ELECTROMAGNETIC RADIATION THROUGH MATTER

When a beam of electromagnetic radiation passes through matter it is *attenuated* and the intensity falls. If the initial intensity is $I_0$, the intensity $I$ after the beam has passed through a slab of thickness $x$ is given by the exponential law (see Chapter 1).

$$I(x) = I_0 e^{-\mu x}, \tag{2.26}$$

where $\mu$ is the *linear attenuation coefficient* and is a characteristic of the material of the slab. This equation can also be written as

$$I(x) = I_0 \exp \left[ - \left( \frac{\mu}{\rho} \right) (\rho x) \right]$$

(2.27)

where $\rho$ is the density of the material and $\mu/\rho$ is the *mass attenuation coefficient*.

The attenuation of the beam must arise from scattering or interaction of the radiation with the material. The scattering from free or nearly free electrons is known as *Thomson scattering*. The incident radiation is assumed to set each electron into forced resonant oscillation such that the electron re-emits radiation of the same frequency but in all directions. If the wavelength of the radiation is very much less than atomic dimensions, each of the $Z$ electrons in an atom scatter independently and the total intensity from an atom is $Z$ times the intensity scattered by a single atom. The scattering by bound electrons, i.e. by the atom as a whole, is known as *Rayleigh scattering* and has the important property that the average radiated power is proportional to $\lambda^{-4}$. In this case there is a coherent sum of the amplitudes for scattering from individual electrons and the sum is squared to obtain the intensity.

When the wavelength is of the same order of magnitude as the atomic dimensions the pattern of the scattered intensity gives information about the electron density of the atom. (This is the quantity $P$ defined by eqn (6.25).) When the wavelength is comparable to the regular spacing of atoms in a crystal, the crystal constitutes a three-dimensional grating and X-ray diffraction may be observed.

It can be deduced from Table 2.2 that important contributions to the attenuation come from Compton scattering and the photoelectric effect. The $Z$ dependence of these processes gives rise to considerable differences in the attenuation coefficient for various materials, especially in the X-ray region as shown in Fig. 2.10.

## PHOTONS AND VISION

It is now established that the visual receptors in the eye are stimulated by the absorption of a single photon. The absorption process gives a signal to the retina, but these signals are not all transmitted individually to the brain. Instead, it is thought that the retina has a data-collecting and scaling function. The rate of receiving quanta in each receptor varies from about one per hour in the dark to hundreds per second in bright sunlight.

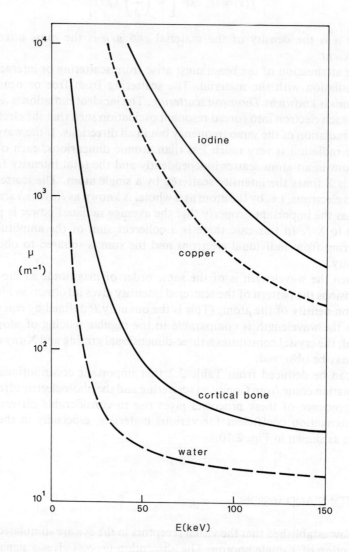

*Figure 2.10.* Variation of the linear attenuation coefficient $\mu$ at X-ray energies for various materials.

## DUAL NATURE OF ELECTROMAGNETIC RADIATION

Radiation from all sections of the electromagnetic spectrum can be made to show wave properties. The interference and diffraction phenomena in the visible region are familiar, and the diffraction of X-rays has important practical applications which are described in Chapter 8. However, in this chapter we have seen that there are a number of phenomena in which the photon concept is required to explain the behaviour of electromagnetic radiation, namely the emission of thermal radiation by hot bodies, the photoelectric effect and the short-wavelength limit of the X-ray spectrum. In addition, in the explanations of Compton scattering and pair production it is necessary to assign particle properties such as momentum to the photon.

Thus it appears that the transfer of energy by radiation occurs in discrete quanta of energy, i.e. discrete and separate energy packets with a definite momentum, but the propagation of radiation is still adequately described by the wave theory. The interpretation of this wave–particle duality is discussed in Chapter 6.

## PROBLEMS

**2.1.** Calculate the quantum of energy for the following cases, giving the answers in joules and in electron-volts:

(*i*)   Infrared radiation of wavelength 0.3 mm.

(*ii*)   X-ray radiation of wavelength 0.3 nm.

(*iii*)   A simple pendulum oscillating with a period of 1 s.

*Solution.*   (*i*)   $E = 6.63 \times 10^{-22}$ J $= 4.15 \times 10^{-3}$ eV
        (*ii*)   $E = 6.63 \times 10^{-16}$ J $= 4.15$ keV
        (*iii*)   $E = 6.63 \times 10^{-34}$ J $= 4.15 \times 10^{-15}$ eV

**2.2.** The work function of a particular photoelectric emitter is 2.0 eV, and light of wavelength 300 nm is used to cause emission. Find the stopping potential, the kinetic energy of the most energetic electrons, and the velocity of these electrons.

*Solution.*

$$\lambda = 300 \text{ nm} = 3.0 \times 10^{-7} \text{ m}$$

$$h\nu = \frac{hc}{\lambda} = \frac{6.625 \times 10^{-34} \times 3 \times 10^{8}}{3 \times 10^{-7}}$$

$$= 6.625 \times 10^{-19} \text{ J}$$

$$= \frac{6.625 \times 10^{-19}}{1.6 \times 10^{-19}} \text{ eV} = 4.14 \text{ eV}.$$

$$V_s e = h\nu - W$$

$$= 4.14 - 2.0 \text{ eV} = 2.14 \text{ eV}.$$

$$\therefore V_s = 2.14 \text{ V}.$$

$$\tfrac{1}{2} m v_{max}^2 = V_s e$$

$$v_{max} = \left\{ 2V_s \frac{e}{m} \right\}^{\frac{1}{2}} = \{2 \times 2.14 \times 1.76 \times 10^{11}\}^{\frac{1}{2}}$$

$$= 8.7 \times 10^5 \text{ m s}^{-1}.$$

**2.3.** Repeat the previous problem for a wavelength of 700 nm. How do you interpret the results in this case?

**2.4.** Calculate the short-wavelength limit for the continuous X-ray radiation emitted by a tungsten target bombarded with 100 keV electrons.

*Solution.*   $\lambda_{min} = 0.012$ nm

**\*2.5.** By considering only the phenomena discussed in this chapter devise a set of experiments whose results could be combined to yield a value for $e/m$ for the electron. Would this be an accurate method of determining this quantity?

**2.6.** A beam of X-rays of wavelength 0.01 nm is incident on a carbon target. The scattered X-rays are detected at an angle of 60° to the direction of the incident beam. Find the wavelength of the scattered X-rays, and the momentum and angle of recoil of the scattered electron.

*Solution.*   $\lambda = 0.0112$ nm, $p_e = 6.35 \times 10^{23}$ kg m s$^{-1}$, $\theta = 53°$

**\*2.7.** A photon of energy $h\nu$ is Compton scattered through an angle $\phi$. Show that the ratio of the kinetic energy of the recoil electron to the energy of the photon is given by

$$\frac{x(1 - \cos \phi)}{1 + x(1 - \cos \phi)}$$

where $x = h\nu/m_0 c^2$.

*Solution.* From eqn (2.19) we have

$$h\nu + m_0c^2 = h\nu' + mc^2$$
$$= h\nu' + T + m_0c^2.$$

$$\therefore T = h\nu - h\nu'$$
$$= \frac{hc}{\lambda} - \frac{hc}{\lambda'}.$$

But from eqn (2.21)

$$\lambda' = \lambda + \frac{h}{m_0c}(1 - \cos\phi)$$

$$\frac{\lambda'}{\lambda} = 1 + \frac{h\nu}{m_0c^2}(1 - \cos\phi) = 1 + x(1 - \cos\phi).$$

Hence

$$T = \frac{hc}{\lambda}\left(1 - \frac{\lambda}{\lambda'}\right)$$
$$= h\nu\left[1 - \frac{1}{1 + x(1 - \cos\phi)}\right]$$

or

$$\frac{T}{h\nu} = \frac{x(1 - \cos\phi)}{1 + x(1 - \cos\phi)}.$$

**2.8.** A positron is slowed down by collisions until it is almost at rest. It then interacts with an electron, also nearly at rest, and the electron–positron pair are annihilated. If the electron is not attached to an atom, find the minimum number of photons which can be produced, their energies and the angle between them.

# 3 | Electrons and Ions

## ELECTROLYSIS

The experiments of Faraday on the electrolysis of solutions of various chemical compounds gave the first evidence of the atomic nature of electricity, and established the laws of electrolysis (1833). These laws can be stated as follows:

(*i*)   The mass of substance liberated as a result of the chemical action of an electric current is proportional to the quantity of electricity which passes through the solution.

(*ii*)   The masses of substances deposited by the passage of the same quantity of electricity are proportional to their chemical equivalent weights.

From these results Faraday inferred that the same amount of electricity is associated with an atom of a given element during electrolysis. He called the charged atom an *ion*. This inference clearly implies that electricity is carried in discrete amounts. In 1874, Stoney advanced the atomic picture by suggesting that there is a natural unit of electricity and proposed the name, *electron*, for this unit.

Faraday's laws can be expressed quantitatively in the form

$$M = QW/Fv, \tag{3.1}$$

where $M$ is the mass liberated, $Q$ is the quantity of electricity or charge passed, $W$ is the atomic weight and $v$ is the valency or number of charges on the ion, so that the equivalent weight is $W/v$. The constant of proportionality $F$ is called the faraday and is the charge which liberates the equivalent weight in kilograms. Equation (3.1) can be used to obtain a value for the *specific charge* of the ions, i.e. the ratio of the charge $q$ to the mass $m$. If a total of $n$ ions is deposited the total charge is $Q = nq$ and the total mass deposited is $M = nm$, and hence

$$\frac{q}{m} = \frac{Q}{M} = \frac{Fv}{W}. \tag{3.2}$$

## CATHODE RAYS

The study of electrolysis revealed the existence of units of positive and negative charge in the form of ions. The existence of units of negative charge characterized by a much larger value of specific charge was first revealed by the study of the conduction of electricity through rarefied gases which led to the discovery of cathode rays. These cathode rays were so called because they emanated from the cathode, and are now known to be emitted as a result of bombardment of the cathode by positive ions formed in the gas. The early experiments revealed the following properties of the cathode rays:

(*i*)   They travel in straight lines.

(*ii*)   They penetrate small thicknesses of material such as metal foils.

(*iii*)   They carry considerable amounts of kinetic energy.

(*iv*)   They are emitted perpendicular to the surface of the cathode.

(*v*)   They carry negative charge.

(*vi*)   They are deflected by electric and magnetic fields.

Some of these properties are consistent with a wave nature for the rays, but properties (v) and (vi) can only be consistent with an interpretation of the rays as negatively charged particles. This corpuscular nature was firmly established by J. J. Thomson (1897), who used the property of deflection in electric and magnetic fields to measure the specific charge.

## MOTION OF CHARGED PARTICLES IN ELECTRIC AND MAGNETIC FIELDS

In order to understand how the specific charge of cathode rays and other particles is measured it is necessary to examine the general behaviour of charged particles in electric and magnetic fields.

(*a*)   *Electric field*.   A particle of charge $q$ at rest in an electric field of intensity $E$ experiences as force

$$F = qE \tag{3.3}$$

and hence acquires an acceleration

$$a = qE/m, \tag{3.4}$$

where $m$ is the mass of the particle. Therefore, a particle initially at rest between two plane parallel condenser plates will be drawn to the positive or negative plate depending on whether the sign of its charge is negative or positive. If the distance between the plates is $d$ and potential difference is $V$, the electric intensity is

$$E = V/d; \qquad (3.5)$$

if the charge moves through the whole distance $d$ the force is $qV/d$ and the work done is $Fd = qV$. The particle acquires kinetic energy and the gain in kinetic energy can be equated to the work done, i.e.

$$\tfrac{1}{2}mv^2 = qV. \qquad (3.6)$$

If there is an aperture in the plate to which the particles are attracted, some of the particles will pass through the aperture and proceed as a beam with constant velocity $v$. This is the principle of the *electron gun*.

When a moving particle enters a region where an electric field acts perpendicular to its motion, the particle is deflected from its initial direction, as shown in Fig. 3.1. The electric field acts in the $y$-direction, so that there is no force in the $x$-direction and the component of velocity $v_x$ is unchanged. If we let the position co-ordinates at time $t$ be $(x, y)$, then the time of flight is given by $t = x/v_x$. The acceleration in the $y$-direction is

$$a_y = \frac{qE}{m} = \frac{qV}{md},$$

so that the deflection in the $y$-direction at time $t$ is given by

$$y = \tfrac{1}{2}a_y t^2 = \frac{qV}{2md}\left(\frac{x}{v_x}\right)^2. \qquad (3.7)$$

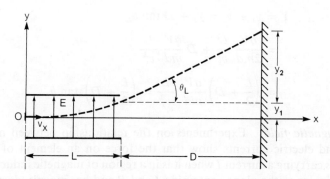

*Figure 3.1.* The deflection of charged particles due to an electric field across a parallel plate condenser.

Thus the relation between the co-ordinates is of the form $y = cx^2$ and the trajectory, in the region where the electric field acts, is a parabola. At any point along this trajectory the direction of motion is given by

$$\tan \theta = \frac{dy}{dx}$$

$$= \frac{qV}{2md}\frac{2x}{v_x^2} = \frac{2y}{x}. \tag{3.8}$$

The deflection $y_1$ at the point when the particle leaves the region where the field acts is obtained by substituting $x = L$, where $L$ is the length of the condenser plates, into eqn (3.7), i.e.

$$y_1 = \frac{qV}{2md}\frac{L^2}{v_x^2},$$

and the direction of motion at this point is obtained by substituting $x = L$, $y = y_1$, into eqn (3.8):

$$\tan \theta_L = \left(\frac{dy}{dx}\right)_{x=L}$$

$$= \frac{2y_1}{L} = \frac{qV}{md}\frac{L}{v_x^2}.$$

It should be noted that the eqns (3.7) and (3.8) are valid in the region where the electric field acts, i.e. $0 \leqslant x \leqslant L$, $0 \leqslant t \leqslant L/v_x$. After leaving the condenser, the particle enters a field-free region and continues in a straight line with constant velocity and direction of motion $\tan \theta_L$. Hence the total deflection $Y$ on a screen, distance $D$ from the end of the condenser, is given by

$$Y = y_1 + y_2 = y_1 + D \tan \theta_L$$

$$= \frac{qV}{2md}\frac{L^2}{v_x^2} + D\frac{qV}{md}\frac{L}{v_x^2}$$

$$= \left(\frac{L}{2} + D\right)\frac{qV}{md}\frac{L}{v_x^2} = \left(\frac{L}{2} + D\right)\tan \theta_L. \tag{3.9}$$

(b)  *Magnetic field.*  Experiments on the relationship between magnetic fields and electric currents show that the force on an element of wire of length $ds$ carrying a current $I$ when it is in a region of magnetic induction $B$ is perpendicular to the plane containing $I$ and $B$ and has magnitude given by

$$F = I \, ds \, B \sin \theta,$$

where $\theta$ is the angle between the direction of the current and the magnetic field. For a particle of charge $q$ and velocity $v$ the current through an element of path $ds$ is $I = qv/ds$, so that the force on the particle in a magnetic field is

$$F = qv B \sin \theta. \tag{3.10}$$

Thus, if $v$ is parallel to $B$ there is no force and if $v$ is zero there is no force. The force is perpendicular to the direction of motion and the magnetic field, and has its maximum value when these are perpendicular to each other. These properties can be summarized by writing eqn (3.10) in vector notation as

$$F = q v \wedge B. \tag{3.11}$$

In the general case, it is necessary to resolve the velocity $v$ into components parallel and perpendicular to the field $B$. In the special case when $\theta = 90°$ there is no component of velocity parallel to $B$. The force always acts perpendicular to the path of the particle so that the tangential velocity is unchanged and the particle moves in a circle. The normal acceleration, directed towards the centre of the circle, is given by

$$a = v^2/R$$

and hence we may equate

$$qvB = ma = mv^2/R \tag{3.12}$$

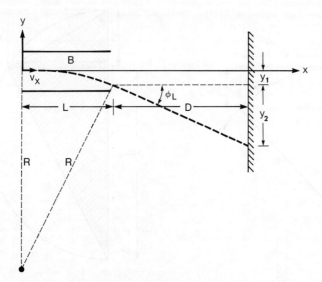

*Figure 3.2.* The deflection of charged particles due to a magnetic field which acts perpendicular to the plane of the diagram.

$$\therefore R = \frac{mv}{qB}. \tag{3.13}$$

It should be noted that $R$ depends on the momentum $p = mv$. Because of this, eqn (3.13) remains valid even when the velocity of the particle is so high that non-relativistic mechanics is no longer valid.

In many cases we are interested in the deflection after the particle has covered a small arc of a circle and then emerged from the region where the magnetic field acts, as shown in Fig. 3.2. The deflection $y$ inside the field may be determined using the diagram given in Fig. 3.3; from the two similar (shaded) triangles we have

$$\frac{y}{x} = \frac{x}{2R - y},$$

and substituting for $R$ from eqn (3.13) this becomes

$$\frac{x^2 + y^2}{2y} = \frac{mv}{qB}. \tag{3.14}$$

The total deflection on a screen at a distance $D$ from the edge of the field is given by

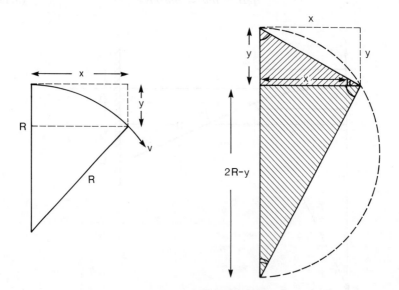

*Figure 3.3.* Geometrical construction drawn to determine the relation between the deflection $y$ in a magnetic field, the distance $x$ travelled in the original direction, and the radius $R$ of the circular motion.

$$Y = y_1 + y_2 = y_1 + D \tan \phi_L$$

where $L$ is the length of the region over which the magnetic field acts, $y_1$ is the value of the deflection given by eqn (3.14) when $x = L$, and $\tan \phi_L$ is the value of the derivative $dy/dx$ at $x = L$.

In the special case when the deflection is sufficiently small that the term $y^2$ in eqn (3.14) can be neglected, we have

$$y = \frac{qB}{2mv} x^2 \qquad (3.15)$$

$$\frac{dy}{dx} = \frac{qB}{mv} x$$

so that the trajectory in the magnetic field approximates to a parabola if the deflection is small. Also,

$$y_1 = \frac{qB}{2mv} L^2, \quad \tan \phi_L = \frac{qB}{mv} L,$$

so that the total deflection becomes

$$Y = \frac{qB}{2mv} L^2 + D \frac{qB}{mv} L$$

$$= \left(\frac{L}{2} + D\right) \frac{qB}{mv} L = \left(\frac{L}{2} + D\right) \tan \phi_L. \qquad (3.16)$$

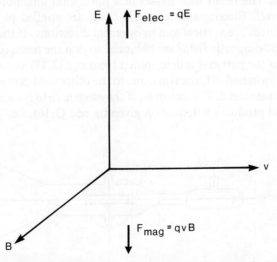

*Figure 3.4.* The direction of the forces on a positively charged particle moving in crossed electric and magnetic fields.

(c)  *Electric and magnetic fields at right angles.*   If electric and magnetic fields act at right angles to each other and to the direction of motion of the particle the forces on the particle act in opposite directions as shown in Fig. 3.4. Thus, the total force on the particle is

$$F = qE - qvB,$$

and for one particular value of the velocity

$$v = E/B \qquad (3.17)$$

the force $F$ is zero. If a beam of particles with a range of velocities is passed through these *crossed fields*, and the electric field is applied across a long, thin condenser so that deflected particles hit the side of the condenser, only those particles whose velocities satisfy eqn (3.17) will emerge. Hence this arrangement acts as a *velocity selector*.

If the particles have a negative charge, the directions of both the electric and magnetic forces are reversed, so that the direction of the deflection due to either field is reversed but the condition for no deflection in crossed fields in unchanged.

## MEASUREMENT OF THE SPECIFIC CHARGE *e/m* FOR ELECTRONS

A version of Thomson's apparatus is shown in Fig. 3.5. A beam of cathode rays passes from the cathode through the two slits which serve to define a narrow beam. The beam then passes in a horizontal direction through the evacuated vessel. Electric and magnetic fields are applied so that the corresponding forces are vertical and in opposite directions. If the strengths of the electric and magnetic fields are balanced so that the beam is undeflected, the velocity of the particles is determined from eqn (3.17) as $v = E/B$. Then if one field is switched off, the force due to the other field causes a deflection which can be measured. For example, if the electric field is switched off, the magnetic field produces a deflection given by eqn (3.16), i.e.

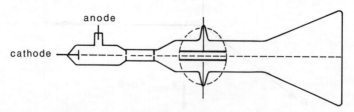

*Figure 3.5.*   A version of J. J. Thomson's apparatus for measurement of *e/m*.

$$y = \frac{qBL}{mv}\left(\frac{L}{2} + D\right).$$ (3.18)

Hence, by rearranging this expression and substituting for $v$, we find the specific charge

$$\frac{q}{m} = \frac{yE}{B^2 L(\frac{1}{2}L + D)}.$$ (3.19)

Thomson's measurements showed that the specific charge of the cathode ray particles is nearly 2000 times greater than the specific charge for the hydrogen ion, which indicated very clearly that these negatively charged particles were very different from the positive ions studied by electrolysis. Thomson made the assumption that the magnitude of the charge of the cathode ray particles is the same as that of a singly charged ion. This unit of charge is called the *electronic charge e*, and the cathode ray particles are called *electrons*. The modern value of the specific charge $e/m$ for electrons is $1.76 \times 10^{11}$ C kg$^{-1}$.

It was found that the value of $e/m$ obtained was independent of the nature of the cathode and also independent of the nature of the positive ions which caused the emission from the cathode. It was also found that the same value of $e/m$ was obtained for the negatively charged particles emitted by the following mechanisms:

(*i*) Spontaneous emission from radioactive atoms. This is the process of $\beta$-decay discussed in Chapter 7.

(*ii*) Thermionic emission from heated metal or oxide filaments.

(*iii*) Photoelectric emission from metals irradiated by short-wavelength light.

(*iv*) Emission from metals due to the action of strong electric fields, described in Chapter 6.

Thus, it may be concluded that *electrons are fundamental constituents of matter*.

## THE ELECTRONIC CHARGE e

The interpretation of the measurement of $e/m$ depends on the assumption that the charge on the electron is the same in magnitude as the charge on a singly ionized atom. Many attempts were made to measure the electronic charge independently, but most of these were inaccurate owing to the

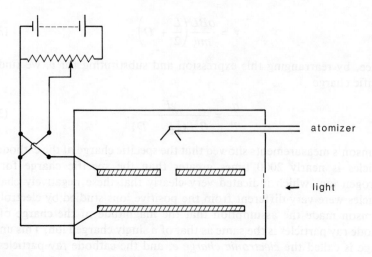

*Figure 3.6.* A version of Millikan's apparatus for measurement of the electronic charge *e*.

difficulty of determining the number of electrons present with any accuracy, and others were indirect and required an accurate knowledge of other constants such as Avogadro's number. These problems were overcome in the experiment devised by Millikan (1909) in which it is possible to make measurements on a small number of electrons.

A version of Millikan's apparatus is shown in Fig. 3.6. Small droplets of a suitable oil are produced by an atomizer jet above a hole in the upper of the two condenser plates. Usually these drops acquire a charge by picking up stray electrons, but an X-ray beam can be used to produce further ionization if necessary. The motion of a single drop is observed through a short-focus telescope; the drop is illuminated from the side and so appears bright due to scattered light. If there is no voltage applied across the condenser plates the drop falls under the influence of the gravitational field until it reaches a constant terminal velocity when the forces on the drop balance. The fall of the drop is then timed over a measured distance so that this terminal velocity can be determined. The forces acting on the drop are drawn in Fig. 3.7a. We let the mass and radius of the drop be $m$ and $a$, respectively, and the density of the oil and of air be $\rho$ and $\rho_a$, respectively. Then the weight of the drop is given by

$$W = mg = \frac{4}{3} \pi a^3 \rho g.$$

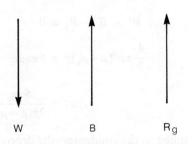

(a) drop falling

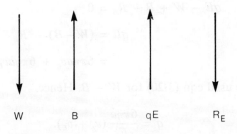

(b) drop rising in electric field

*Figure 3.7.* The forces of an oil drop in Millikan's experiment when (a) the drop is falling and there is no electric field and (b) the drop is rising in an electric field.

and the buoyancy or upthrust is given by Archimedes' principle as

$$B = \frac{4}{3}\pi a^3 \rho_a g.$$

The resistance of the air is always in opposition to the motion and is given by Stoke's law as

$$R_g = 6\pi \eta a v_g,$$

where $\eta$ is the viscosity of the air and $v_g$ is the terminal velocity. At the terminal velocity these forces balance so that

$$W - B - R_g = 0$$

$$\frac{4}{3}\pi a^3(\rho - \rho_a)g = 6\pi\eta a v_g \tag{3.20}$$

$$\therefore a^2 = \frac{9\eta v_g}{2g(\rho - \rho_a)}. \tag{3.21}$$

If a voltage is applied to the condenser, the drop can be made to rise or fall, depending on the strength of the electric field and its direction. We consider the case when the drop rises and reaches a terminal velocity $v_E$. The forces on the drop are as shown in Fig. 3.7b; the weight and buoyancy are unchanged in magnitude and direction, but the resistance of the air changes direction in order to oppose the motion and its magnitude now depends on $v_E$. When the forces balance we have

$$qE - W + B - R_E = 0$$

$$qE = (W - B) + R_E$$

$$= 6\pi\eta a v_g + 6\pi\eta a v_E,$$

where we have used eqn (3.20) for $W - B$. Hence

$$q = \frac{6\pi\eta a}{E}(v_g + v_E). \tag{3.22}$$

Since the radius $a$ can be determined from eqn (3.21), the charge $q$ on the drop can be determined from eqn (3.22).

In the experiment, a single drop is timed over measured distances, rising and falling, for as long as possible, and these measurements are repeated with other drops. It is found that the charge on the drop is always an integral multiple of a certain basic value, i.e.

$$q = ne, \tag{3.23}$$

where $n$ is an integer, and the charge on a drop always changes by a discrete amount which is again an integral multiple of the basic unit $e$. Thus, this experiment gives direct proof that electric charge always occurs in discrete amounts which are integral multiples of the electronic charge $e$, and so establishes the discreteness or atomicity of charge. It also establishes the electron as the *fundamental unit of charge*, as had been proposed by Stoney much earlier.

## THE VARIATION OF ELECTRON MASS WITH VELOCITY

The theory of special relativity predicts that the mass of a particle increases with velocity according to the relation given in eqn (2.7). No such dependence is predicted for the electronic charge which is taken to be a universal constant independent of the velocity of the charge. Thus the specific charge $e/m$ of an electron of velocity $v$ can be written as

$$\frac{e}{m} = \frac{e}{m_0}\left(1 - \frac{v^2}{c^2}\right)^{\frac{1}{2}} \tag{3.24}$$

where $m_0$ is the rest mass. The velocities of the electrons produced in Thomson's cathode ray tube were very much less than $c$ so that Thomson and his successors effectively measured the ratio $e/m_0$, but it is clear that a measurement of $e/m$ for electrons with higher velocities can be used to verify the variation of mass with velocity.

## POSITIVE RAYS

If the cathode of a cathode ray tube or discharge tube is perforated, the positive ions pass through and continue moving in the direction of the electric field. These positive ions were first called positive rays. Measurement of their specific charge $q/m$ showed that its value is very much less than $e/m$ for electrons and depends on the nature of the gas in the tube. This establishes clearly that the positive rays were indeed positive ions which are formed by ionization in the gas and then drawn to the cathode by the electric field.

## MEASUREMENT OF $q/m$ FOR POSITIVE IONS

J. J. Thomson also devised a method of measuring the specific charge of positive ions. A fine beam of positive ions is produced and these pass through a region where an electric field $E$ and a magnetic field $B$ act in the same direction. If we let the direction of the fields be the $y$-direction and the direction of the initial motion of the ions be the $x$-direction, as shown in Fig. 3.8, the electric field gives rise to a deflection in the $y$-direction whose magnitude can be obtained from eqn (3.9) as

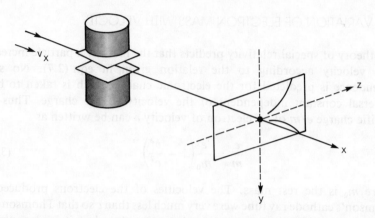

*Figure 3.8.* The formation of parabolas in the measurement of $q/m$ for positive ions. The magnetic field $B$ acts in the positive $y$-direction. The parabola in the lower quadrant is obtained if the electric field is in the same direction, and the parabola in the upper quadrant is obtained if $E$ is reversed.

$$y = k_1 \frac{q}{m} \frac{E}{v_x^2}, \tag{3.25a}$$

whereas the magnetic field leads to a deflection in the $z$-direction whose magnitude is given by eqn (3.16) as

$$z = k_2 \frac{q}{m} \frac{B}{v_x}, \tag{3.25b}$$

where $k_1$ and $k_2$ are constants depending on the dimensions of the system. Combining these equations to eliminate the velocity $v_x$, we find

$$z^2 = \frac{k_2^2}{k_1} \frac{q}{m} \frac{B^2}{E} y, \tag{3.26}$$

which is the equation of a parabola. Thus, ions with the same value of $q/m$ strike the photographic plate at a series of points along a parabola, each point corresponding to a different value of $v_x$, while for a different value of $q/m$ a different parabola is observed.

Modern instruments for the measurement of $q/m$ usually display the presence of various ions as lines on a photographic plate. These instruments are called *mass spectrographs* and the photographic record is called the *mass spectrum*.

## ISOTOPES

The measurements of $q/m$ for positive ions indicated that atoms of the same chemical element could have different masses. This phenomenon had already been observed in the study of transformations of radioactive atoms, and the term *isotopes* had been introduced by Soddy (1913) to describe atoms with the same chemical properties but different mass. The existence of isotopes is now known to be due to the different composition of the nuclei of atoms.

## ISOTOPIC AND ATOMIC WEIGHTS

The masses of atoms and ions can be quoted in absolute units in terms of the kilogram. It is convenient, however, to quote the isotopic masses in terms of a relative unit, known as the *atomic mass unit* (amu). This unit is now defined to be such that the mass of the most abundant isotope of carbon is exactly 12 mass units. It is then found that the masses of all other isotopes are very nearly integer numbers of amu; this integer is known as the *mass number A* and is used to characterize the different isotopes of the same element. The actual element is characterized by the chemical symbol or by the atomic number $Z$ which denotes the position in the periodic table. Thus the isotopes of carbon can be written as shown in Table 3.1.

The *atomic weights* determined from the laws of chemical combination depend on the abundance of the various stable isotopes present. Earlier chemical measurements took oxygen as the standard, but it has recently been agreed to use the $^{12}C$ scale for chemical and physical measurements.

*Table 3.1.* Isotopes of carbon ($Z = 6$)

| Isotopes | Mass (amu) | Abundance |
|----------|------------|-----------|
| $^9C$    |            | Not stable |
| $^{10}C$ | 10.0168    | Not stable |
| $^{11}C$ | 11.0114    | Not stable |
| $^{12}C$ | 12.0000    | 98.89%    |
| $^{13}C$ | 13.0034    | 1.11%     |
| $^{14}C$ | 14.0032    | Not stable |
| $^{15}C$ | 15.0106    | Not stable |

## WAVE–PARTICLE DUALITY

It was found that a satisfactory description of Compton scattering (see Chapter 2) could be given by introducing the concept of the photon and attributing to it a momentum

$$p = \frac{h\nu}{c} = \frac{h}{\lambda}.$$

It was suggested by de Broglie (1924) that the dual behaviour should apply also to material objects, which should show wave-like characteristics associated with a wavelength $\lambda$ given by

$$\lambda = \frac{h}{p} = \frac{h}{mv}. \tag{3.27}$$

Now, if we insert into eqn (3.27) typical values of $m$ and $v$ for laboratory-scale objects, for example $m = 1$ g, $v = 30$ m s$^{-1}$, we find that $\lambda$ is excessively small,

$$\lambda = \frac{6.625 \times 10^{-34}}{10^{-3} \times 30} = 2.2 \times 10^{-32} \text{ m}.$$

Thus, in order to have an observable wavelength the magnitude of the momentum must be very small, and this means that the mass must be small, for otherwise the motion would be imperceptible and no experiment could be carried out in a reasonable length of time. For an electron with velocity $10^7$ m s$^{-1}$ we have

$$\lambda = 0.7 \times 10^{-10} \text{ m} = 0.07 \text{ nm}.$$

This wavelength is similar to that for hard X-rays and, therefore, it should be possible to demonstrate the wavelike characteristics of electrons by diffracting an electron beam using a crystal as a diffraction grating, as described in Chapter 8.

Experiments by Davisson and Germer (1927) and by G. P. Thomson (1928) first demonstrated that diffraction patterns could be observed with an electron beam by reflection from a single crystal or by transmission through a thin film. Figure 3.9 shows the patterns observed by transmission of X-ray and electron beams through thin samples, and it is clear that these patterns are the same in nature. G. P. Thomson used the Debye–Scherrer method, which gives these ring patterns, to study electron diffraction, while Davisson and Germer originally used the Bragg spectrometer method. By using a substance whose crystal spacing is known it is possible to determine the electron wavelength experimentally and to check that it agrees with that predicted by

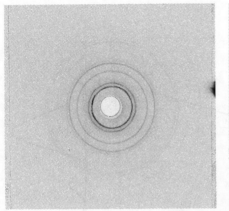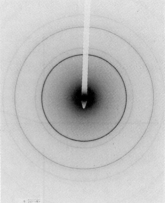

*Figure 3.9.* Ring patterns produced by diffraction of radiation from small crystals randomly oriented (Debye–Scherrer method). The picture on the left shows the result of electron diffraction by an aluminium foil and the picture on the right shows the result of X-ray diffraction by small crystals of aluminium. (Pictures are not to the same scale.) (Courtesy Dr D.C. Povey, Department of Chemistry, and Mr J. Greaves, Department of Materials Science and Engineering, University of Surrey.)

the de Broglie relation (3.27). Similarly, Fig. 8.10 shows that comparable diffraction patterns are given by X-rays and neutrons.

We may conclude that *there is an inherent wave–particle duality in nature* and that those objects which we have hitherto regarded as "particles" can under suitable circumstances show wave properties. It is unfortunate that the language we must use to describe atomic phenomena stems from our everyday experience, in which there is normally a sharp division between particle and wave phenomena, but it must now be accepted that the pictorial representation which these terms imply is valid only for classical, laboratory-scale systems.

## THE UNCERTAINTY PRINCIPLE

The wave–particle duality places a limitation on the precision with which we can determine all the properties of a particle. As an example, we consider the diffraction of a beam of electrons by a single slit, as shown in Fig. 3.10, and assume that the width of the slit and the initial momentum of the beam are known. The resulting diffraction pattern is recorded on a screen or

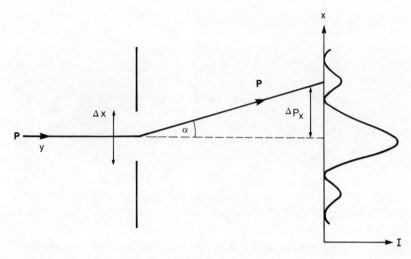

*Figure 3.10.*   The diffraction of an electron beam by a single slit.

photographic plate, and the variation in intensity is as shown in the figure. The fact that a diffraction pattern has been observed implies that the electrons have indeed passed through the slit and have exhibited their wave properties. The fact that they have passed through the slit means that the position of the electrons perpendicular to the direction of the beam is specified to this extent and we can take the width of the slit as a measure of the uncertainty $\Delta x$ in position. On the other hand, the fact that there is a diffraction pattern which extends outside the geometrical shadow region implies that an electron acquires a component of momentum $\Delta p_x$ in the $x$-direction. The resultant momentum must be unchanged, so that $\Delta p_x$ is given by

$$\Delta p_x = p \sin \alpha. \tag{3.28}$$

The wave theory for diffraction gives the relation

$$\Delta x \sin \alpha = c\lambda$$

where the values $c = 1, 2, 3, \ldots$ determine the values of the $\sin \alpha$ corresponding to the minima in the diffraction pattern, and $c = 1.5$ gives the value of $\sin \alpha$ for the first subsidiary maximum. For simplicity we write

$$\Delta x \sin \alpha \approx \lambda, \tag{3.29}$$

and substituting this into eqn (3.28) we have

$$\Delta x \, \Delta p_x \approx p\lambda,$$

which, using eqn (3.27), becomes

$$\Delta x \, \Delta p_x \approx h. \tag{3.30}$$

Thus, as a result of defining the position of the electron within an uncertainty $\Delta x$, the electron has acquired an uncertainty in momentum in the same direction of $\Delta p_x$, and the product of these two uncertainties is of the order of Planck's constant.

Equation (3.30) is a mathematical expression of *Heisenberg's uncertainty principle* which states that *it is fundamentally impossible to make simultaneous measurements on the momentum and position to an accuracy better than one quantum of action h*. Many other examples have been studied which always lead to the same conclusion. The uncertainty principle is not limited to measurements of momentum and position but applies to many pairs of conjugate variables, the product of whose dimensions yield the dimensions of action (J s). It should be noted, however, that the uncertainty principle applies to position and momentum measured in the same direction, i.e. to $\Delta x \, \Delta p_x$, etc.

As a further example of the uncertainty principle, we consider the emission and absorption of radiation by an atom. We assume that an atom has been raised to an excited state of energy $W_n$ by some means, such as electron bombardment. If the atom makes a transition to a lower state of energy $W_m$, the emitted radiation has a well-defined energy given by

$$E = h\nu = W_n - W_m.$$

This means that the emitted radiation is monochromatic, i.e. it consists of a single frequency $\nu$ and can be represented by a sine wave. If we now use this radiation to excite another atom of the same type, the whole photon energy $h\nu$ is transferred and we know the energy of the excited atom exactly, but because the radiation is represented by a sine wave of infinite length we have no possibility of measuring the time at which the excitation occurred. In order to measure the time we must construct a wave packet of length $l$ so that the uncertainty in the time is reduced to $\Delta t = l/c$. An expression (derived in Chapter 6) for the length of a wave packet is given by

$$l = 2\pi/\Delta k$$

so that

$$\Delta t = \frac{l}{c} = \frac{2\pi}{c \, \Delta k}, \tag{3.31}$$

where $\Delta k$ is the spread in the wave numbers of the sine waves which have been used to construct the wave packet. For electromagnetic waves, we have $k = 2\pi/\lambda$ and $E = h\nu$, so that

$$k = \frac{2\pi E}{hc}$$

$$\Delta k \approx \frac{2\pi}{hc} \Delta E,$$

and substituting for $\Delta k$ in eqn (3.31) we have

$$\Delta t \, \Delta E \approx h.$$

In order to measure the time, it is necessary to introduce other frequencies in order to make a wave packet, and this immediately destroys the exactness of the energy measurement.

Because of the importance of the quantity $(h/2\pi)$ (see Chapters 6 and 7) it is now more customary to write

$$\Delta x \, \Delta p_x \gtrsim (h/2\pi) \tag{3.32a}$$

$$\Delta y \, \Delta p_y \gtrsim (h/2\pi) \tag{3.32b}$$

$$\Delta z \, \Delta p_z \gtrsim (h/2\pi) \tag{3.32c}$$

$$\Delta t \, \Delta E \gtrsim (h/2\pi) \tag{3.32d}$$

$$\Delta \phi \, \Delta J_z \gtrsim (h/2\pi) \tag{3.32e}$$

where $E$ is the energy of a particle and $t$ is the time at which it is measured, $J_z$ is the component of angular momentum of a particle in the $z$-direction and $\phi$ is its angular position in the $xy$-plane.

## THE PRINCIPLE OF COMPLEMENTARITY

One of the important consequences of the uncertainty principle is that it is not possible to determine wave and particle properties exactly at the same time. This means that an attempt to determine the particle properties exactly eliminates any possibility of simultaneously observing any aspect of the wave-like properties, and vice versa. An example of this situation would be a two-slit electron interference experiment in which we attempted to define the electron trajectory either by closing one slit or by placing some detecting apparatus immediately behind one of the slits. This procedure would destroy the interference pattern. It follows that any experiment which can be devised displays *either* the particle-like *or* the wave-like characteristics of the system. This suggests that the wave and particle pictures give complementary descriptions of the same system.

## PROBLEMS

**3.1.** A current of 1 A is passed through an electrolyte for 1 h. Given that the faraday $F$ is equal to 96 500 C, show that the mass deposited is equal to 0.0373 times the equivalent weight.

**\*3.2.** Particles of mass $m$ and charge $q$ are injected at a velocity $v$ along the axis of a long parallel-plate condenser which has potential $V$ applied across it. Show that if no particles emerge from the far end of the condenser, the ratio $L/d$ of the length $L$ to the separation $d$ of the condenser plates must be greater than $v(m/qV)^{\frac{1}{2}}$. Check that the quantity $v(m/qV)^{\frac{1}{2}}$ is dimensionless.

**3.3.** A positive ion with velocity $v$ in the horizontal direction is undeflected when it passes through crossed electric and magnetic fields of intensity $2000$ V m$^{-1}$ and $10^{-3}$ T. Find the velocity $v$. If the ratio of the charge to the mass of the ion is $9.65 \times 10^7$ C kg$^{-1}$, find the ratio of the force on the ion due to gravity and the force due to the electric field, and also the ratio of the force due to gravity and the force due to the magnetic field. Comment on the results.

*Solution.* The forces on the ion are $qE$, $qvB$ and $mg$. Since the first two balance we have

$$v = \frac{E}{B} = \frac{2 \times 10^3}{10^{-3}} = 2 \times 10^6 \text{ m s}^{-1}.$$

Also, the ratios $mg/qE$ and $mg/qvB$ must be equal. Hence

$$\frac{mg}{qE} = \frac{mg}{qvB} = \frac{9.81}{9.65 \times 10^7 \times 2 \times 10^3} = 0.51 \times 10^{-10}.$$

Thus, the force due to gravity is completely negligible.

**3.4.** Electrons emitted by a heated filament are accelerated across a potential difference of 200 V. Find the final velocity of the electrons. Repeat the calculation for singly charged $^{12}$C ions. (Note: 1 amu is equivalent to $1.660 \times 10^{-27}$ kg.)

*Solution.* $v = 0.84 \times 10^7$ m s$^{-1}$ (electron), $v = 0.58 \times 10^5$ m s$^{-1}$ (C ion).

**3.5.** Electrons with velocity $v = 2 \times 10^7$ m s$^{-1}$ are injected along the axis of parallel-plate condenser of length $L = 1.6 \times 10^{-2}$ m and separation $d = 5 \times 10^{-3}$ m. If the potential difference across the condenser is 50 V find the deflection and direction of motion of the electrons as they leave the condenser. If a screen is placed a distance $D = 0.20$ m away from the end of the condenser find the total deflection on the screen.

*Solution.* The deflection at the end of the condenser is obtained by substituting $x = L$ into eqn (3.7):

$$y_1 = \frac{eV}{2md} \frac{L^2}{v_x^2}$$

$$= \frac{1.76 \times 10^{11} \times 50}{2 \times 0.5 \times 10^{-2}} \frac{(1.6)^2}{4 \times 10^{14}} \times 10^{-4}$$

$$= 5.6 \times 10^{-4} \text{ m.}$$

The direction of motion is obtained by substituting $x = L$, $y = y_1$ into eqn (3.8):

$$\tan \theta_L = \frac{2y_1}{L} = \frac{2 \times 5.6 \times 10^{-4}}{1.6 \times 10^{-2}} = 0.07$$

$$\theta_L \approx 4°.$$

Total deflection

$$= y_1 + D \tan \theta_L$$

$$= 5.6 \times 10^{-4} + 20 \times 10^{-2} \times 0.07 \text{ m}$$

$$= 1.46 \times 10^{-2} \text{ m.}$$

**3.6.** Electrons of velocity $v = 2 \times 10^7$ m s$^{-1}$ are passed into a magnetic field which acts over a distance of $1.6 \times 10^{-2}$ m, and the total deflection is observed on a screen at a distance of $0.20$ m away from the edge of the field. Find the magnitude and direction of the field required to produce a deflection of $1.46 \times 10^{-2}$ m on the screen.

*Solution.*  $B = 0.88 \times 10^{-3}$ T

**\*3.7.** Two ions of charge $q$ and mass $m_1$ and $m_2$, respectively, pass through a velocity selector under the influence of crossed electric and magnetic fields. They then travel through a semicircle under the influence of the magnetic field alone and strike a photographic plate. Show that the radius of the semicircle is a linear function of the charge to mass ratio, and that this system can therefore be used to measure isotopic masses.

**3.8.** A charged oil drop falls a distance of $5 \times 10^{-3}$ m in 20 s at its terminal velocity and in the absence of an electric field. If the density of the oil is

$0.80 \times 10^3$ kg m$^{-3}$ and that of the air is 1.20 kg m$^{-3}$, and the viscosity of air is $1.81 \times 10^{-5}$ N s m$^{-2}$, find the mass and radius of the drop. If an electric field of $2 \times 10^5$ V m$^{-1}$ is applied and the drop carries two electronic charges, find the ratio of the force on the drop due to gravity and the force due to the electric field. Why is the ratio so different from that found in Problem 3.3?

*Solution.*   $m = 1.57 \times 10^{-14}$ kg, $a = 1.68 \times 10^{-6}$ m, ratio of forces = 2.4

**3.9.**   Find the de Broglie wavelength for the following particles and discuss the feasibility of observing diffraction in each case:

*(a)*   a pellet of mass 1.1 g and velocity $3 \times 10^2$ m s$^{-1}$;

*(b)*   a student of mass 70 kg and velocity 5 m s$^{-1}$;

*(c)*   an electron with velocity $10^7$ m s$^{-1}$;

*(d)*   a pellet of mass 1 g and velocity $10^{-20}$ m s$^{-1}$.

*Solution.*
*(a)*   $\lambda = 2.01 \times 10^{-33}$ m
*(b)*   $\lambda = 1.89 \times 10^{-35}$ m
*(c)*   $\lambda = 0.73 \times 10^{-10}$ m
*(d)*   $\lambda = 0.66 \times 10^{-10}$ m

**3.10.**   A column of people walking at normal pace in single file pass through a doorway into a large room. Determine what the magnitude of Planck's constant would need to be for the doorway to act as a diffraction slit. What would be observed as a result of this diffraction effect?

*Solution.*   $\lambda \approx$ width of doorway, $h \approx 10^2$–$10^3$ J s

**3.11.**   Show that the de Broglie wavelength in ångström units for an electron accelerated from rest through a potential difference of $V$ volts is given

*(a)*   non-relativisitically, by

$$\lambda = 12.27 \ V^{-\frac{1}{2}},$$

*(b)*   relativistically by

$$\lambda = 12.27 \ V^{-\frac{1}{2}} \left( 1 + \frac{Ve}{2m_0c^2} \right)^{\frac{1}{2}}.$$

*Solution.*   *(b)* In this case the kinetic energy is $T = eV$, as in case *(a)*, but $T \neq \frac{1}{2}mv^2$, so that we cannot find the momentum directly from $T$. Instead we must use the relativistic formulae. We have

$$E^2 = p^2c^2 + m_0^2c^4, \quad E = T + m_0c^2,$$
$$T = eV$$

so that

$$p^2c^2 = E^2 - m_0^2c^4 = (T + m_0c^2)^2 - m_0^2c^4$$
$$= T^2 + 2m_0c^2T$$
$$p^2 = 2m_0T\left(1 + \frac{T}{2m_0c^2}\right)$$
$$p^2 = \sqrt{2m_0eV}\left(1 + \frac{eV}{2m_0c^2}\right)^{\frac{1}{2}}$$
$$\lambda = \frac{h}{p} = \frac{h}{\sqrt{2m_0e}}\, V^{-\frac{1}{2}}\left(1 + \frac{eV}{2m_0c^2}\right)^{-\frac{1}{2}}$$

# 4 | Models of the Atom

## ELECTRIC CHARGES IN ATOMS

The experiments described in Chapter 3 established the electron charge as the fundamental unit of electricity, and established the electron itself as a fundamental constituent of matter. By convention, the sign of the charge on the electron is taken to be negative. Positive charge is carried by ions which are essentially different in nature to electrons; their masses are determined by the nature of the substances which has been ionized and are very much greater than the electron mass. Negative ions can also be produced as a result of ionizing a molecule, but ionization of an atom always leads to the emission of electrons and never to positively charged fundamental particles. Thus, we may observe the ionization processes

$$NaCl \rightarrow Na^+ + Cl^-$$
$$Na \rightarrow Na^+ + e^-$$

but we *do not* observe the ionization process

$$Cl \rightarrow Cl^- + e^+.$$

(As we have seen in Chapter 2, there is a positive electron $e^+$ but its creation requires an amount of energy far in excess of that available in ionization processes.) The next stage in our study of the atom is, therefore, to seek information about the nature of the positively charged part of the atom, and the magnitude of its charge. With this information it will be possible to make models of the atom, by making assumptions about the way that the positively charged part of the atom and the constituent electrons are arranged to form a stable atom.

## THE PRELIMINARY EXPERIMENTS

The earliest information about the structure of the atom came from the experiments by Lenard between 1895 and 1903. He studied the collisions of an electron beam as it passed through a gas, and found that fast electrons can pass through the gas without noticeable deviations in their paths. He also observed that, when deflections did occur, for a fixed electron energy the amount of scattering was proportional to the mass of the gas atoms. From measurements on the decrease of intensity of such a beam, due to the scattering of the electrons as they pass through the gas, it is possible to determine the effective cross-sectional area presented to the incident beam by each atom, and from this to deduce an atomic radius, as explained in Chapter 1.

Lenard found that at low electron energies the effective atomic radius was in agreement with the value of approximately $10^{-10}$ m calculated from the kinetic theory, but as the electron energy was raised the effective radius decreased towards a limiting value of about $10^{-14}$ m.

The evidence from Lenard's experiments was confirmed and extended by the experiments of Rutherford, Geiger and Marsden (1909–1913) on the scattering of $\alpha$-particles. These $\alpha$-particles are the *nuclei* of helium atoms and have a positive charge of $+2e$ units. They are spontaneously ejected from the atoms of certain heavy elements, such as thorium and uranium, with energies of the order of a few MeV. Rutherford and his collaborators found that even with this much larger and more massive projectile, the majority of the $\alpha$-particles in the beam experienced little or no deflection, and the number of $\alpha$-particles which experienced small-angle deflections was consistent with a simple theory of scattering. On the other hand, the number of $\alpha$-particles which experienced large-angle deflections, though few, was very much in excess of the number predicted from the law which predicted the small-angle scattering. For example, in one of Geiger's experiments on scattering from a thin gold foil, the mean angle of deflection was found to be $1°$. The law for small-angle scattering predicts that the number of $\alpha$-particles scattered through an angle equal to or greater than $\theta$ is given by

$$N(\theta) = N_0 e^{-(\theta/\theta_m)^2}, \tag{4.1}$$

where $\theta_m$ is the mean deflection and $N_0$ is the initial number of $\alpha$-particles. Hence, for $\theta = 10°$ and $\theta_m = 1°$, we have

$$N(10°) = N_0 e^{-100} \approx 10^{-43} N_0,$$

and for $\theta = 90°$ we have

$$N(90°) = N_0 e^{-8100} \approx 10^{-3500} N_0,$$

but Geiger found that the number scattered through angles of $90°$ or greater was

$$N(90°) \approx 1.2 \times 10^{-4} N_0.$$

The discrepancy between the prediction and the observation is so great that the mechanisms of small-angle scattering and large-angle scattering must be totally different.

## SOME SIMPLE MODELS

In 1907, J. J. Thomson proposed a model of the atom which has since become known as the "plum pudding" or "currant bun" model. It was assumed that the positive charge is distributed uniformly over a sphere whose radius is equal to the observed atomic radius of approximately $10^{-10}$ m. The electrons must be distributed in this sphere in such a way that the whole system is stable and electrically neutral. This model will obviously explain the ionization process. Thomson was also able to show that this model will yield the small-angle scattering law (4.1). Since the electrons are arranged in the positive part of the Thomson atom in such a way as to make the system stable, the variation in the net charge in any part of the atom is small, and hence there can be only slight variations in the force on the projectile as it passes through or near the "currant bun" atom. This means that the scattering of $\alpha$-particles passing through a thin foil composed of such atoms is due to a large number of small deflections randomly directed. This model, however, cannot provide any explanation for the importance of large-angle scattering.

From the results of his experiment, Lenard drew the conclusion that atoms do not appear like impenetrable massive spheres, at least when bombarded by fast projectiles, but should be thought of as consisting partly of empty space. He introduced a picture of the atom consisting of a number of very small impenetrable centres surrounded by a loose cloud of electrons extending out to a radius of about $10^{-10}$ m. Slow projectiles would be scattered by the electron cloud and so would measure the full radius of the atom. Fast or heavy projectiles would experience little or no scattering as they pass through the electron cloud but would experience a large deflection when they impinge on one of the impenetrable centres. Thus, this picture yields a qualitative explanation, not only of both large-angle and small-angle scattering, but also of the variation of the effective atomic radius with the energy of the projectile.

## RUTHERFORD'S MODEL

An accurate and quantitative description of large-angle scattering was derived by Rutherford, who assumed that the positive charge of the atom is

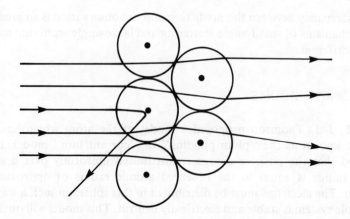

*Figure 4.1.* Large- and small-angle scattering of $\alpha$-particles from an array of atoms.

concentrated in a single impenetrable centre, called the *nucleus* of the atom, and surrounded by the electron cloud. The qualitative description of the scattering of $\alpha$-particles from such an atom is the same as in Lenard's model and is illustrated in Fig. 4.1. The large-scale scattering arises from the electrostatic repulsion between the $\alpha$-particle of charge $+2e$ and the nucleus of charge $+Ze$. The electrostatic force is given by Coulomb's law as

$$F = \frac{2Ze^2}{4\pi\epsilon_0 r^2}, \tag{4.2}$$

and, since $Z \approx \frac{1}{2}A$, where $A$ is the atomic mass number defined in Chapter 3, the force between the $\alpha$-particle and the nucleus can be very strong at short distances and even sufficient to turn the $\alpha$-particle back through 180°, as shown in Fig. 4.2. The application of classical mechanics to the problem of the scattering of the $\alpha$-particle due to the force $F$ yields the result that the number of $\alpha$-particles scattered through an angle lying between $\theta$ and $\theta + d\theta$ is

$$N(\theta) = N_0\, ns \frac{Z^2 e^4}{(4\pi\epsilon_0)^2 4 T_\alpha^2} \frac{1}{\sin^4 \frac{1}{2}\theta}\, d\Omega, \tag{4.3}$$

where $n$ is the number of atoms per unit volume of the target, $s$ is the thickness of the target, $d\Omega = 2\pi \sin\theta$, $N_0$ is the number of $\alpha$-particles in the incident beam, and $T_\alpha = \frac{1}{2}M_\alpha v^2$ is the kinetic energy of the $\alpha$-particles.

The validity of the formula (4.3) was tested extensively by Geiger and

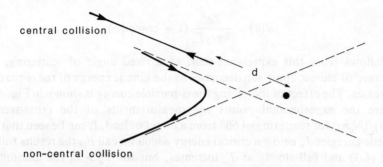

*Figure 4.2.* The scattering of α-particles from a single nucleus showing a non-central collision and a central collision.

Marsden, and it has been tested many times since for different types of projectiles. It is always found to be valid provided that the kinetic energy of the projectile does not exceed a critical value (see below). Thus, the experiments by Rutherford and his collaborators establish the nuclear model of the atom. The evidence from the large-angle scattering of α-particles was indisputable; it was, as Rutherford remarked, "almost as incredible as if you had fired a 15-inch shell at a piece of tissue paper and it came right back and hit you".

## THE NUCLEUS

In Fig. 4.2 the distance *d* denotes the *distance of closest approach* or apsidal distance for a head-on or central collision. As the α-particle approaches the nucleus it is slowed up until, at the point where it is turned back from the nucleus, all the kinetic energy has been converted into potential energy. Hence

$$T_\alpha = \frac{2Ze^2}{4\pi\epsilon_0 d},$$

i.e.

$$d = \frac{Ze^2}{2\pi\epsilon_0 T_\alpha} = \frac{Ze^2}{\pi\epsilon_0 M_\alpha v^2}. \qquad (4.4)$$

For a non-central collision which leads to scattering through an angle θ, the distance of closest approach is defined as the distance between the vertex of the path of the projectile and the nucleus, and is given by

$$d(\theta) = \frac{Ze^2}{4\pi\epsilon_0 T_\alpha} (1 + \mathrm{cosec}\, \tfrac{1}{2}\theta) \qquad (4.5)$$

It follows from this expression that, at a fixed angle of scattering, the distance of closest approach decreases as the kinetic energy of the $\alpha$-particle increases. The effect of increasing the $\alpha$-particle energy is shown in Fig. 4.3, where the experimental points are measurements of the *cross-section* $N(\theta)/(N_0 ns)$ for scattering of 60° from a target of lead. It can be seen that for kinetic energies $T_\alpha$ below a critical energy which we call $E_0$, the results follow eqn (4.3) and fall slowly as $T_\alpha$ increases, but for $T_\alpha > E_0$ the amount of scattering is substantially changed. The variation of the apsidal distance or distance of closest approach with the energy $T_\alpha$ is also shown by the lower dashed curve in the figure, and from this it can be seen that the critical energy corresponds to a distance $d \approx 1.3 \times 10^{-14}$ m. The apparatus used for these measurements is also shown in the figure.

It appears from the results shown in Fig. 4.3 that, when the $\alpha$-particle approaches as close as $10^{-14}$ m to the centre of the atom, the predictions derived from Coulomb's law for the interaction between the $\alpha$-particle and the nucleus are no longer valid. This is taken as evidence for the existence of an additional force acting between the $\alpha$-particle and the nucleus. Since the $\alpha$-particle is itself a nucleus, this force must arise from the interaction between the two nuclei, and we therefore call it the *nuclear force*. At distances very much greater than $10^{-14}$ m, the $\alpha$-particle is sensitive to the presence of the nucleus only through the Coulomb force, i.e. the nucleus appears like a point charge, but at distances of the order of $10^{-14}$ m the $\alpha$-particle is sensitive to the specifically nuclear properties of the nucleus through the nuclear force. We may, therefore, take the distance of closest approach at the critical energy as a very rough estimate of the size of the nucleus, and if we assume that the nucleus is a uniform sphere of constant density, we may determine the density of the lead nucleus. We have

radius $\approx$ distance of closest approach at $E_0$

$$\approx 1.3 \times 10^{-14} \text{ m} \qquad (4.6)$$

---

*Figure 4.3.* (a) Apparatus used to study $\alpha$-particle scattering for nuclei. (b) The relative cross-sections for $\alpha$-particle scattering at 60° from a lead target plotted as a function of $\alpha$-particle energy. The critical energy is the energy at which the experimental results depart from the Coulomb curve. The lower dashed curve and scale on the right of the diagram shows the distance of closest approach as a function of $\alpha$-particle energy. (Courtesy of The American Physical Society. From G. W. Farwell and H. E. Wegner, *Phys. Rev.* **95**, 1212, 1954.)

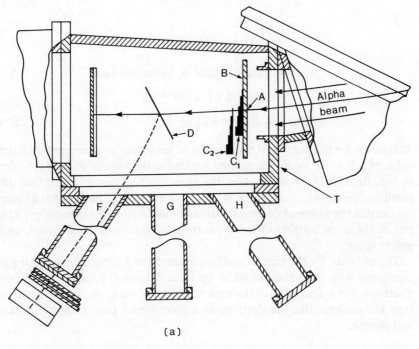

(a)

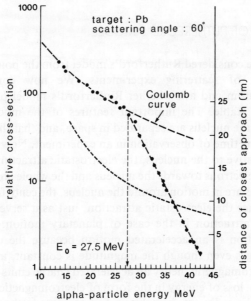

target : Pb
scattering angle : 60°

Coulomb curve

$E_0$ = 27.5 MeV

(b)

$$\text{volume} \approx \frac{4\pi}{3} (1.3 \times 10^{-14})^3 \text{ m}^3$$

$$\text{mass} \approx \text{atomic weight} \times \text{hydrogen mass}$$

$$\approx 208 \times 1.67 \times 10^{-27} \text{ kg}$$

$$\text{density} \approx 0.4 \times 10^{17} \text{ kg m}^{-3}. \tag{4.7}$$

This result for the density of the nucleus of lead must be compared with the value of $1.1 \times 10^4$ kg m$^{-3}$ measured for lead in the form of a metal cylinder or bar. Because the whole mass of the atom has been concentrated into the small nucleus at the centre of the atom, the density of the nucleus has a value far outside the range of densities normally measured in the laboratory. The rest of the atom consists of a few electrons of almost negligible mass, and empty space.

The estimate for the nuclear radius obtained from $\alpha$-particle scattering is consistent with the value obtained by Lenard from the scattering of fast electrons. For a discussion of the basic structure of the atom it is sufficient to treat the nucleus, like the electron, as a point object possessing both mass and charge.

## LIMITATIONS OF RUTHERFORD'S MODEL

So far we have considered Rutherford's model from the point of view of the interpretation of scattering experiments. We now consider a single undisturbed atom and ask whether Rutherford's model predicts that this atom will be stable. The important features of the model are that the electrons and the nucleus are separated in space, and that they remain so, at least during the time of observation in an experiment. Now, if the electrons are at rest relative to the nucleus, the electrostatic attraction of the nucleus will draw the electrons towards the nucleus and the whole atom will collapse. If the electrons are in motion around the nucleus, the centrifugal force might serve to balance the electrostatic attraction, just as it serves to balance the gravitational attraction in the case of planetary motion. Unfortunately, uniform rotation is an accelerated motion because the direction of the velocity changes even though the magnitude is constant, and according to classical electromagnetic theory an accelerated charge emits electromagnetic radiation. The loss of energy in the form of electromagnetic energy causes a reduction in the radius of the orbit and an increase in the velocity and the acceleration of the electron (see Problem 4.2). Thus the electron would spiral into the nucleus while emitting a continuous spectrum of electromagnetic radiation. The time taken for this process has been estimated to be of the order of $10^{-8}$ s.

It appears that the Rutherford model, at least in the simple form we have used so far, predicts that the atom is not stable and that it emits a continuous spectrum of radiation. However, we have already seen in Chapter 2 that some of the classical theory of electromagnetic radiation must be modified in order to describe interaction of radiation with atoms. We therefore examine the information available on the emission of radiation by atoms in order to find a way in which Rutherford's model can be applied to stable atoms.

## THE BASIC FEATURES OF ATOMIC SPECTRA

With the aid of a suitable spectrometer it is possible to separate and identify the wavelengths emitted by a variety of sources. These studies show that the spectrum of radiation emitted by individual atoms consists of a discrete set of wavelengths, and that the spectrum is characteristic of the element concerned. This phenomenon was used as early as 1860 by Kirchhoff and Bunsen to identify new elements, and it has developed into a powerful tool in chemical analysis, the simplest form of which is the so-called "flame test" used in elementary chemistry.

The spectrum seen with the aid of a spectrometer appears as a set of images of the spectrometer slit. Each slit image corresponds to one of the wavelengths in the spectrum and appears on a photographic plate as a fairly sharp dark line*, as shown in Fig. 4.4. These slit images are known as *spectral lines* and the whole spectrum is known as a *line spectrum*. Other types of spectra are observed, such as the continuous spectra of thermal radiation from hot bodies discussed in Chapter 2 and the band spectra emitted by molecules, but only the atomic line spectra will be discussed here.

In this chapter we consider in detail only the spectrum of the simplest atom, namely the hydrogen atom. The spectral lines of hydrogen which appear in the visible and the near ultraviolet regions are listed in Table 4.1 and plotted in Fig. 4.5. It can be seen that the lines crowd together in the near ultraviolet region and converge towards a short-wavelength limit. This is a characteristic feature of spectral lines observed in all atoms, although the position of the limit depends on the type of atom. Many attempts were made to find some numerical relationship between the magnitudes of the wavelengths of this series of hydrogen lines; eventually Balmer (1885) was able to

---

* If viewed directly through the spectrometer, the lines would appear bright against a dark background, but the effect of the radiation on the photographic plate causes darkening.

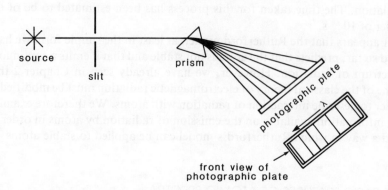

*Figure 4.4.* Schematic diagram of a prism spectrometer showing how the dispersive effect of the prism can be used to study atomic spectra. A modern instrument would use a grating instead of a prism to produce the dispersion.

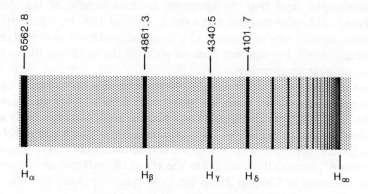

*Figure 4.5.* The emission spectrum of atomic hydrogen.

show that the reciprocal of the wavelength was given to a high degree of accuracy by the formula

$$\frac{1}{\lambda} = R_{\mathrm{H}} \left( \frac{1}{2^2} - \frac{1}{n^2} \right), \tag{4.8}$$

and the series has become known as the Balmer series. The values of the integer $n$ are given in Table 4.1, and $R_{\mathrm{H}}$ is a constant, known as the *Rydberg constant*, for hydrogen. Modern measurements of this constant yield a value of $10\ 967\ 757 \pm 1.2\ \mathrm{m}^{-1}$. Subsequently, other series of hydrogen lines were

*Table 4.1.* Observed wavelengths in the Balmer series of hydrogen

| Symbol for line | $n$ | Wavelength (Å) |
|---|---|---|
| $H_\alpha$ | 3 | 6562.85 |
| $H_\beta$ | 4 | 4861.30 |
| $H_\gamma$ | 5 | 4340.47 |
| $H_\delta$ | 6 | 4101.73 |
| $H_\epsilon$ | 7 | 3970.07 |
| . | | |
| . | | |
| . | | |
| . | 14 | 3722.04 |
| . | 15 | 3712.14 |

discovered in the ultraviolet and infrared regions, and they all obey the general formula

$$\frac{1}{\lambda} = R_H \left( \frac{1}{m^2} - \frac{1}{n^2} \right), \quad n > m, \qquad (4.9)$$

where the integer $m$ is fixed for a particular series and $n$ is a variable integer which runs from $m + 1$ to $\infty$. At the moment, eqn (4.8) represents nothing more than a convenient though somewhat curious mathematical formula. We note, however, that it can be converted into an energy formula by using Planck's expression for the energy of a photon. Thus

$$h\nu = \frac{hc}{\lambda} = hc\, R_H \left( \frac{1}{m^2} - \frac{1}{n^2} \right)$$

and we can rewrite this equation in the form

$$h\nu_{nm} = W_n - W_m, \quad n > m \qquad (4.10)$$

so that the energy of the photon of wavelength $\lambda$ emitted by an atom is given by the difference between two energies $W_n$ and $W_m$, where

$$W_m = -hc\, R_H/m^2, \qquad (4.11)$$

and similarly for $W_n$. If we label the photon frequency by the appropriate integers, the photon energy corresponding to the difference between the energy terms $W_s$ and $W_m$ is given by

$$h\nu_{sm} = W_s - W_m.$$

If we now add and subtract $W_n$ on the right-hand side of this equation, we have

$$h\nu_{sm} = W_s - W_n + W_n - W_m$$

$$\therefore h\nu_{sm} = h\nu_{sn} + h\nu_{nm}, \quad s > n, n > m, \tag{4.12}$$

The result embodied in eqn (4.12) was originally discovered empirically by Ritz (1908), who called it the *combination principle*. Not all possible combinations predicted by eqns (4.10) and (4.12) are observed, as there are so-called *selection rules* which eliminate certain combinations.

The fundamental importance of eqns (4.10) and (4.12) is that it is possible to represent the energy of photons emitted by an atom to produce a given spectral line as the difference between two energy terms each labelled by an integer, and that by addition and subtraction of these energy terms it is possible to calculate the energy and wavelength of other lines. The inclusion of the minus sign in the definition of $W_m$ and the rearrangement of terms in eqns (4.10) is not essential to this argument but appear at this stage as an arbitrary choice which must be justified later.

## BOHR'S QUANTUM POSTULATES

The great step forward in the understanding of the hydrogen spectrum and its interpretation in terms of Rutherford's model was made by Niels Bohr in 1913. Drawing on the work of Planck and Einstein, described in Chapter 2, which showed that radiation is emitted and absorbed in discrete amounts and which established the concept of the photon, Bohr assumed that the classical electromagnetic theory was not completely valid for atomic systems. His assumptions can be stated in the form of two postulates:

(*i*)   An atom can exist in certain allowed or stationary states each of which has a definite energy. When the atom is in one of these states it is stable and does not radiate.

(*ii*)   The emission or absorption of radiation by an atom is associated with a transition between two stationary states. The radiation is emitted or absorbed as a single quantum whose energy is determined by the difference in energy of the two states.

It follows from the second postulate that the energy equation for emission of radiation is

$$h\nu = W_n - W_m, \tag{4.13}$$

where $\nu$ is the frequency of the radiation, $W_n$ and $W_m$ are the energies of the initial and final states, and $W_n > W_m$. Thus, Bohr's postulates are completely consistent with the eqns (4.10)–(4.12) deduced from a study of the experimental data on atomic spectra, and they are also consistent with the concept of quantization of electromagnetic radiation. The new feature introduced in the postulates is the concept of *quantization* of the energy of atomic states, since the atom is assumed to exist only in certain states each having a definite energy. Once this feature is introduced the complex behaviour of atomic spectra becomes comprehensible, and the interpretation through differences in energy terms, which can now be associated with the energies of stationary states, becomes meaningful.

It should be noted that the description of the allowed atomic states as "stationary" does not mean that the electrons are at rest. However, it is implicit that the electrons are bound since the atom is stable and not necessarily ionized. If we consider the hydrogen atom and take the zero of energy when the electron is just free with zero kinetic energy then it can be seen that the energies $W_n$ and $W_m$ are binding energies and must be negative. An amount of energy $+ |W_m|$ must be supplied to the atom to release an electron from the $m$th state. This explains our choice of sign in eqn (4.11); it remains to be seen whether we can prove that the energies of the states of the hydrogen atom have the magnitudes given by that expression.

## THE PLANETARY MODEL

We have as yet no way of calculating the energies of the stationary states from first principles, but Bohr was able to show how this could be done for a simple planetary model of the hydrogen atom. The model is a simple extension of Rutherford's model, in which it is assumed that the single electron is separated from the nucleus and moves in a circle of radius $r$ around the nucleus. For simplicity we assume that the hydrogen nucleus is infinitely heavy compared with the electron and can, therefore, be considered to be at rest at the centre of the orbit. (This restriction can easily be removed; see below and Problem 4.6.) We also follow Bohr's assumption that the motion of the electron in a stationary state can be described using classical mechanics, although it is far from obvious that this is a good assumption as we have already assumed that classical electrodynamics is not valid in the stationary state.

In order to give the formulae slightly greater generality, we write the nuclear charge as $Ze$, even though $Z = 1$ for hydrogen. The equation for the circular motion becomes

$$\frac{mv^2}{r} = \frac{Ze^2}{4\pi\epsilon_0 r^2}, \tag{4.14}$$

so that the kinetic energy of the electron is

$$T = \tfrac{1}{2}mv^2 = \frac{Ze^2}{8\pi\epsilon_0}. \tag{4.15}$$

The potential energy $V$ and total energy $W$ of the electron are given by

$$V = -\frac{Ze^2}{4\pi\epsilon_0 r} \tag{4.16}$$

$$W = T + V = -\frac{Ze^2}{8\pi\epsilon_0 r}. \tag{4.17}$$

The first postulate implies that the energy $W$ can take only certain discrete values. Bohr found that this quantization could be introduced by imposing the condition that the angular momentum of the electron is given by

$$mvr = n\frac{h}{2\pi} \tag{4.18}$$

where $n$ is any integer excluding zero. Combining eqns (4.14) and (4.18) to eliminate the velocity $v$, we find

$$r = n^2 \frac{\epsilon_0 h^2}{\pi m Z e^2}, \tag{4.19}$$

and inserting this expression for $r$ into eqn (4.17) we have

$$W = -\frac{mZ^2e^4}{8\epsilon_0^2 h^2}\frac{1}{n^2}. \tag{4.20}$$

Thus, we see that in the planetary model both the energy of the electron and the radius of its orbit are quantized. The allowed values of these quantities are determined by the integer $n$ which is called a *quantum number*. Later, when we have introduced more quantum numbers, the quantum number $n$ which relates to the energy of the system will be called the *principal quantum number*.

Bohr's second postulate can now be applied to give the energy of the photon emitted in a transition from the stationary state characterized by the quantum number $n$ to the state with quantum number $m$. Hence

$$h\nu = W_n - W_m$$

$$= -\frac{mZ^2e^4}{8\epsilon_0^2h^2}\left(\frac{1}{n^2} - \frac{1}{m^2}\right)$$

$$= \frac{mZ^2e^4}{8\epsilon_0^2h^2}\left(\frac{1}{m^2} - \frac{1}{n^2}\right), \tag{4.21}$$

and the wavelength of the radiation is given by

$$\frac{1}{\lambda} = \frac{h\nu}{hc} = \frac{Z^2me^4}{8\epsilon_0^2h^3c}\left(\frac{1}{m^2} - \frac{1}{n^2}\right). \tag{4.22}$$

When the finite mass $M$ of the nucleus is taken into account (see Problem 4.6) a factor of $\{1 + (m/M)\}^{-1}$ must be included on the right-hand side of eqns (4.19)–(4.22).

We now note that eqns (4.20), (4.21) and (4.22) have exactly the same form as eqns (4.11), (4.10) and (4.9), provided that we put $Z = 1$ and identify the Rydberg constant for hydrogen as

$$R_H = R_\infty \frac{1}{1 + (m/M)}, \tag{4.23}$$

where

$$R_\infty = \frac{me^4}{8\epsilon_0^2h^3c}, \tag{4.24}$$

and since this predicted value of $R_H$ depends entirely on known constants, it serves as an immediate check on the theory. The value of $R_H$ calculated in this way agrees with the experimental value to within 3 parts in $10^5$.

The planetary model and the two postulates constitute what is generally known as the Bohr quantum theory of the atom. As we have seen, it gives results for the hydrogen spectrum which may seem quite astonishingly accurate when we consider the simplicity of the model. It can be applied with similar success to other hydrogen-like atoms which contain only one electron: $He^+$, $Li^{2+}$, etc. Unfortunately, there is no simple way of dealing with second and further electrons in the planetary model. Also, the quantization has been introduced rather arbitrarily through the condition imposed on the angular momentum. For these and other reasons, the planetary model has been superseded by the more fundamental theory described in Chapter 6. The postulates remain valid, however, and have been found to be applicable to all situations involving radiative transitions between states in atoms, molecules or nuclei.

## THE CORRESPONDENCE PRINCIPLE

When the quantum number $n$ becomes very large, successive values of the allowed energies given by eqn (4.20) differ by very small amounts. Thus the allowed energy values appear almost continuous. Similarly, when the spectral lines approach the series limit, they crowd together so that the spectrum appears almost continuous. This behaviour led Bohr to postulate the *correspondence principle* (1923). This principle states that the predictions of the quantum theory for the behaviour of an atomic system must correspond to the predictions of classical theory when the quantum theory is taken to the classical limit. In the present case the classical limit for the quantum theory is the situation of transitions between nearby states having very large quantum number, and it can be shown (see Problem 4.5) that the quantum and classical theories do give identical predictions for the radiated frequency in this limit. Bohr used the correspondence principle to deduce what selection rules apply to transitions between particular states in atoms.

## PROBLEMS

**4.1.** The measured values of the critical energy $E_0$ at which departures from Coulomb scattering of $\alpha$-particles from nuclei occur are given below for several targets. Calculate the distance of closest approach in each case.

| Target element | Scattering angle | $Z$ | $E_0$(MeV) |
|---|---|---|---|
| Silver | 61° | 47 | 16.2 |
| Gold | 60° | 79 | 27.0 |
| Thorium | 60° | 90 | 29.5 |

$(4\pi\epsilon_0 = 1.1 \times 10^{-10}$ F m$^{-1})$

*Solution.* Silver, 12.66 fm; gold, 12.75 fm; thorium, 13.35 fm

**4.2.** Consider an electron moving in a circle around a nucleus of charge $+ Ze$. If the total energy of the electron in an orbit with radius $r_1$ is greater than the total energy in an orbit with radius $r_2$, show that the radius $r_1$ is greater than $r_2$ and also that the velocity and acceleration in the orbit with radius $r_2$ are greater than in the orbit with radius $r_1$.

*Solution.* The formulae for the kinetic energy, potential energy and total energy are given by eqns (4.15)–(4.17), as

$$T = \tfrac{1}{2}mv^2 = \frac{Ze^2}{8\pi\epsilon_0 r} \qquad (4.15)$$

$$V = -\frac{Ze^2}{4\pi\epsilon_0 r} \qquad (4.16)$$

$$W = T + V = -\frac{Ze^2}{8\pi\epsilon_0 r} = -\tfrac{1}{2}mv^2. \qquad (4.17)$$

If $W_1 > W_2$ it follows that

$$W_1 - W_2 > 0$$

$$-\frac{Ze^2}{4\pi\epsilon_0 r_1} + \frac{Ze^2}{4\pi\epsilon_0 r_2} > 0$$

$$\frac{Ze^2}{4\pi\epsilon_0}\left(\frac{1}{r_2} - \frac{1}{r_1}\right) > 0, \quad \text{i.e.} \quad r_1 > r_2.$$

From eqn (4.15) we have

$$\tfrac{1}{2}m(v_2^2 - v_1^2) = \frac{Ze^2}{8\pi\epsilon_0}\left(\frac{1}{r_2} - \frac{1}{r_1}\right)$$

$$\therefore v_2^2 - v_1^2 > 0$$

$$v_2 > v_1.$$

Alternatively, we could have obtained this result using the condition $W_1 - W_2 > 0$ with $W_1 = -\tfrac{1}{2}mv_1^2$ and $W_2 = -\tfrac{1}{2}mv_2^2$.

The force on an electron in a circular orbit is $mv^2/r$ and, hence, the acceleration is $f = v^2/r$. But eqn (4.14) gives

$$\frac{mv^2}{r} = \frac{Ze^2}{4\pi\epsilon_0 r^2},$$

hence

$$f = \frac{Ze^2}{4\pi\epsilon_0 mr^2}$$

$$f_2 - f_1 = \frac{Ze^2}{4\pi\epsilon_0 m}\left(\frac{1}{r_2^2} - \frac{1}{r_1^2}\right)$$

$$\therefore f_2 > f_1.$$

**4.3.** Calculate the velocity of the electron in the orbit $n = 1$ of hydrogen. Hence show that the use of non-relativistic mechanics in the planetary model was justified, to a good approximation. Calculate the radius of the orbit and verify that this is consistent with the values for atomic radii measured in scattering experiments with slow electrons.

*Solution.*   $v = 0.0145c; r = 0.053$ nm

**4.4.** Using the combination principle and the information given in Table 4.1, deduce the wavelength of the hyrogen line emitted as a result of a transition between state $n = 5$ and state $n = 3$.

*Solution.*   $\lambda = 1285$ nm

**4.5.** Show that the angular frequency of revolution of an electron in its orbit is given by the Bohr theory as $w = \pi m Z^2 e^4 / 2\epsilon_0^2 h^3 n^3$. Hence show that when $n$ is very large the frequency of revolution $w/2\pi$ is equal to the frequency of the radiation emitted in the transition of an electron from state $n_2 = n + 1$ to state $n_1 = n$. Comment on the significance of this result.

*Solution.*   From the Bohr theory the angular frequency of revolution is given by

$$mr^2 w = \frac{nh}{2\pi},$$

where

$$r = \frac{n^2 h^2 \epsilon_0}{\pi m e^2 Z}$$

$$\therefore\ w = \frac{nh}{2\pi m} \left( \frac{\pi m e^2 Z}{n^2 h^2 \epsilon_0} \right)^2 = \frac{\pi m Z^2 e^4}{2\epsilon_0^2 h^3 n^3},$$

$$\text{frequency of revolution} = \frac{w}{2\pi} = \frac{m Z^2 e^4}{4\epsilon_0^2 h^3 n^3}.$$

The frequency of emitted radiation for the transition $n + 1 \rightarrow n$ is given by

$$\nu = \frac{m Z^2 e^4}{8\epsilon_0^2 h^3} \left( \frac{1}{n^2} - \frac{1}{(n+1)^2} \right)$$

$$= \frac{m Z^2 e^4}{8\epsilon_0^2 h^3} \frac{2n + 1}{n^2(n+1)^2}.$$

For sufficiently large $n$

$$\frac{2n + 1}{n^2(n + 1)^2} \approx \frac{2}{n^3}$$

and

$$\therefore \nu \approx \frac{mZ^2e^4}{4\epsilon_0^2h^3n^3} \approx \frac{w}{2\pi}.$$

Classical electromagnetic theory predicts that the frequency of emitted radiation is equal to the frequency of revolution, and in the limit of transitions between nearby states having very large quantum numbers the classical theory and the Bohr theory yield the same results. Thus, in this respect the theories satisfy the correspondence principle.

**\*4.6.** Show that the Rydberg constant for hydrogen is given by

$$R_H = R_\infty \frac{1}{1 + (m/M)},$$

where $m$ is the electron mass and $M$ is the proton mass.

*Solution.* Since the nucleus is not infinitely heavy there is no reason to assume that it is at rest. Instead we assume that the electron of mass $m$ and the hydrogen nucleus (the proton) of mass $M$ rotate about their centre of mass with angular frequency $w$. The separation between the electron and proton is still denoted by $r$ and the distances from the centre of mass are denoted by $a$ and $A$, respectively, as shown in the figure.

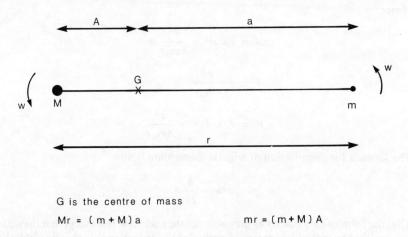

G is the centre of mass

$Mr = (m + M)a$                         $mr = (m + M)A$

The motion of an electron and nucleus about their common centre of mass.

Taking moments we have

$$mr = (m + M)A, \quad \therefore A = \frac{m}{m + M} r$$

$$Mr = (m + M)a, \quad \therefore a = \frac{M}{m + M} r.$$

The total kinetic energy is now

$$\frac{1}{2} MA^2 w^2 + \frac{1}{2} ma^2 w^2$$

$$= \frac{1}{2} r^2 w^2 \left[ \frac{Mm^2}{(m + M)^2} + \frac{mM^2}{(m + M)^2} \right]$$

$$\frac{1}{2} \mu r^2 w^2,$$

where $\mu$ is the *reduced mass* (see eqn (1.2))

$$= \frac{mM}{m + M}.$$

The total angular momentum is now

$$MA^2 W + ma^2 w = \mu r^2 w.$$

The equation of motion of the electron in a circle is now

$$\frac{e^2}{4\pi\epsilon_0 r} = maw^2 = \mu r w^2.$$

Hence

$$T = \frac{1}{2} \mu r w^2 = \frac{e^2}{8\pi\epsilon_0 r}$$

$$V = - \frac{e^2}{4\pi\epsilon_0 r}$$

$$W = T + V = - \frac{e^2}{8\pi\epsilon_0 r}.$$

The formula for quantization of angular momentum is now

$$\mu r^2 w = \frac{h}{2\pi} n.$$

Thus the formulae are exactly as they were for the nucleus at rest except that the mass $m$ is replaced by the reduced mass $\mu$. Continuing the derivation as in the text yields the results (with $Z = 1$)

$$r = \frac{n^2 h^2 \epsilon_0}{\pi \mu e^2}$$

$$W = -\frac{e^4 \mu}{8\epsilon_0^2 n^2 h^2}$$

$$\frac{1}{\lambda} = \frac{e^4 \mu}{8\epsilon_0^2 h^3 c}\left(\frac{1}{n_1^2} - \frac{1}{n_2^2}\right)$$

$$R_H = \frac{e^4 \mu}{8\epsilon_0^2 h^3 c}$$

$$\therefore R_H = R_\infty \frac{\mu}{m} = R_\infty \frac{1}{1 + (m/M)}.$$

**\*4.7.** Using the result of the previous question, show that

$$\frac{m}{M} = \frac{R_{He} - R_H}{R_H - \frac{1}{4}R_{He}},$$

where $R_{He}$ is the Rydberg constant for singly-ionized helium, He$^+$. Compare the formula for the wave number of lines in the Balmer series for hydrogen which ends on $n_1 = 2$ with the formula for the wave number of the lines in the series of He$^+$ which ends on $n_1 = 4$. Hence suggest a method of measuring the difference between $R_H$ and $R_{He}$.

**4.8.** Prove that $h/2\pi$ has units of angular momentum.

# 5 | Experimental Evidence for Quantization

In the previous chapter it was shown that the experimental observations on the line spectra emitted by atoms could not be explained in terms of classical theory. In contrast, application of Bohr's assumption that atomic systems exist in stationary states with quantized energy values led to an adequate explanation of the basic features of the hydrogen spectrum. In this chapter, the experimental evidence for quantization is examined in more detail.

## ENERGY LEVELS

The allowed energy values for a quantized system are often referred to as *energy levels*. It is very useful to plot these energy values, as horizontal lines separated by the appropriate spacings, to obtain an *energy level diagram*. For example, eqn (4.20) gives the energy levels of the hydrogen atom as predicted by the planetary model, and the corresponding energy level diagram is shown in Fig. 5.1. For hydrogen, each line represents a possible energy state or level for the single electron; the lowest allowed state with $n = 1$ is called the *ground state*, while the other discrete energy states are called *excited states*. Transitions between the states are represented by arrows on the diagram. If the electron is raised to a higher state, energy must be applied to the atom, and if the electron then falls back to a lower state, energy is given out in the form of a quantum of radiation whose energy $h\nu$ is determined by eqn (4.21). According to eqn (4.20) the energy of the system is zero when $n = \infty$, i.e. the electron is no longer bound to the atom. Since we have as yet no evidence that the energy of a free electron is quantized, the region on the energy level diagram for $E > 0$ is shaded to indicate a continuous region of unrestricted positive energy values for the free electron.

For other elements the energy level diagrams are more complicated, as can be seen from the energy level diagram for sodium shown in Fig. 5.2. In

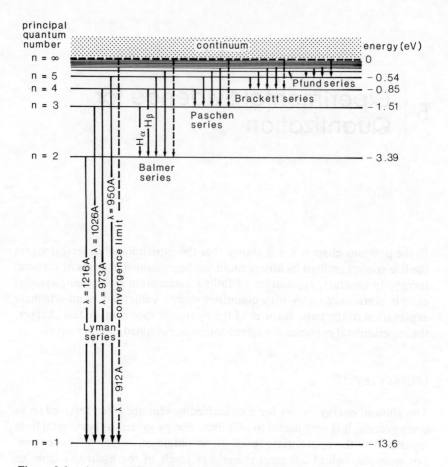

*Figure 5.1.*   An energy level diagram for the hydrogen atom.

hydrogen there is only one electron so that the state of this electron and the state of the atom are synonymous, but in all other atoms the possible states of the atom must be deduced from the experimental data.

## RESONANCE, EXCITATION AND IONIZATION POTENTIALS

The energy necessary to excite an atom from its ground state to a higher state can be supplied in a variety of ways. One very convenient method is by means of electron bombardment. If electrons are emitted from a hot filament and

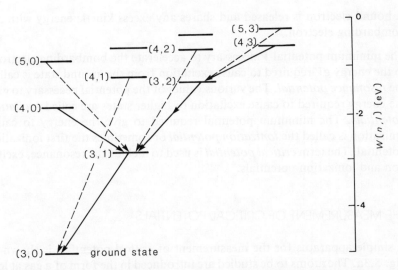

*Figure 5.2.* Some energy levels of the sodium atom labelled with the quantum numbers $(n, l)$. An unbroken arrow indicates the transition leading to the first line in each spectral series while a dashed arrow corresponds to the second line.

accelerated across a potential $V$, they acquire a kinetic energy $T = eV$ which can be transferred to an atom as a result of a collision. We let $W_m$, $W_n$, . . . $W_\infty$ represent the sequence of possible energy states of the atom, where $W_m$ is the ground state, $W_n$ the first excited state, i.e. $n > m$, and so on. The possible processes arising from electron bombardment can then be listed as follows:

(*i*)   If $T < W_n - W_m$, the electron does not carry sufficient energy to excite the atom. There can only be "billiard ball" or *elastic collisions* leading to a change in the kinetic energy of the atom as a whole but no change in its internal energy.

(*ii*)   If $T = W_n - W_m$, the electron has sufficient energy to excite the atom to its next state by transferring energy to one of the bound electrons. This is an *inelastic collision*. The bombarding electron is left with essentially zero kinetic energy. It subsequently executes random thermal motion at an energy which can be determined from the kinetic theory.

(*iii*)   If $T = (W_s - W_m) + dT$, where $s > m$, an inelastic collision can occur leading to excitation of the atom to the state $s$. The free electron carries away the excess kinetic energy $dT$.

(*iv*)   If $T > W_\infty - W_m$, an inelastic collision leading to ionization can occur.

A bound electron is released and shares any excess kinetic energy with the bombarding electron.

The minimum potential $V$ necessary to accelerate the bombarding electrons to the energy $eV$ required to cause excitation from the ground state is called the *resonance potential*. The various values of the potential necessary to give the energy required to cause excitation of higher states are called *excitation potentials*. The minimum potential required to give the energy to cause ionization is called the *ionization potential* or sometimes the first ionization potential. The term *critical potential* is used to include the resonance, excitation and ionization potentials.

## THE MEASUREMENT OF CRITICAL POTENTIALS

A simple apparatus for the measurement of critical potentials is shown in Fig. 5.3a. The atoms to be studied are introduced in the form of a gas at low pressure. The electrons are accelerated from the filament F to the grid G by the potential $V_1$ and meet a small retarding potential $V_2$ between the grid G and the collecting plate P, such that $V_2 \ll V_1$. As $V_1$ is increased from zero up to the resonance potential $V_R$ the collisions of the electrons in the gas are elastic, but when $V_1 = V_R$ the electrons can make inelastic collisions near the grid G. Electrons which do make such inelastic collisions lose all their kinetic energy and are unable to overcome the retarding potential $V_2$, so that the current drops sharply, as shown in Fig. 5.3b. The remaining electrons which have not made inelastic collisions pass through the grid to give a small residual current. As the potential $V_1$ is increased still more, the electrons can make inelastic collisions further in from the grid G and the residual kinetic energy remaining after the collision is sufficient to overcome the retarding potential, so that the current rises again until $V_1 = 2V_R$, when the electrons can make two inelastic collisions before reaching the grid. The pressure of the gas is chosen to make the probability of collision sufficiently high so that there is little possibility for the electrons to reach energies much greater than $eV_R$ before making an inelastic collision. If the gas pressure is reduced, some electrons acquire sufficient energy to excite the higher states of the atom and the position of some of the excitation potentials, and the ionization potentials are seen as bumps on the side of the main curve of current against voltage $V_1$. More elaborate experimental arrangements can be devised to give more accurate determination of the excitation potentials and to distinguish between excitation and ionization potentials.

Apart from the measurement of critical potentials which leads to location of the position of the energy levels, experiments of this type demonstrate the

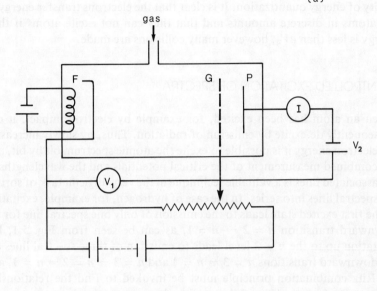

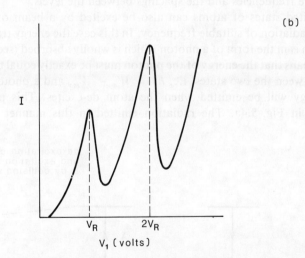

*Figure 5.3.* (a) An apparatus for measurement of the resonance potential.
(b) Results obtained for the current reaching the plate as a function of the voltage
$V_1$ on the grid.

reality of energy quantization. It is clear that the electrons transfer energy to the atoms in discrete amounts and that they can not excite atoms if their energy is less than $eV_R$, however many collisions are made.

## CONTROLLED EXCITATION OF SPECTRA

When an atom has been excited, for example by electron impact, it can subsequently de-excite by emission of radiation. Thus, by slowly increasing the electron energy it is possible to excite the atomic spectrum bit by bit, and the combined measurement of the critical potentials and the wavelengths of the associated lines is a valuable technique in the very difficult task of sorting the spectral lines into series. In the case of hydrogen, for example, excitation of the first excited state leads to the emission of only one spectral line for the downward transition $n = 2 \rightarrow n = 1$, as can be seen from Fig. 5.1, but excitation up to the $n = 3$ level leads to emission of three spectral lines for the downward transitions $n = 3 \rightarrow n = 1$ and $n = 3 \rightarrow n = 2 \rightarrow n = 1$, and the Ritz combination principle must be invoked to find the relationship between the frequencies and the spacings between the levels.

The excited states of atoms can also be excited by a beam of electromagnetic radiation of suitable frequency. In this case the energy transferred to the atom is in the form of a photon which is wholly absorbed (see Chapter 2). This means that the energy of the photon must be exactly equal to the difference between the two states, i.e. $h\nu = W_n - W_m$, and a photon of this same energy will be emitted when the atom de-excites. This process is illustrated in Fig. 5.4a. The radiation emitted in this manner is called

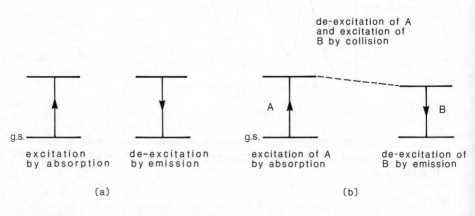

*Figure 5.4.*   The transitions occurring in (a) resonance fluorescence and (b) sensitized fluorescence.

*resonance radiation* and the process is called *resonance fluorescence*.

It is also possible to excite a mixture of two types of atoms, say A and B, by irradiation with a line from the spectrum of type A only. The incident radiation must correspond to a transition involving the ground state of the atoms of type A so that the incident photons can be absorbed to excite these atoms from their ground state to an excited state. If these excited atoms make collisions with atoms of type B they can transfer energy to the latter by exciting them to an appropriate excited state, as shown diagrammatically in Fig. 5.4b. The atoms of type A are left in the ground state and carry away any excess energy in the form of kinetic energy, while the atoms of type B can de-excite by emission of radiation. This process is known as *sensitized fluorescence*.

## PHOTOIONIZATION

When an atom absorbs a photon of sufficiently high energy, ionization can occur and an electron is emitted. The photon energy must be at least equal to the ionization energy, and any excess energy will be carried away by the electron in the form of kinetic energy. This process is completely equivalent to the photoelectric effect in metals, and is sometimes called the atomic photoelectric effect. As a result of photoionization, a beam of high-energy photons passing through matter causes ionization along its path.

## ABSORPTION SPECTRA

When atoms in their ground state are irradiated with a continuous spectrum, certain wavelengths are absorbed, and a record of the spectrum on a photographic plate shows bright lines corresponding to these missing wavelengths, which stand out from the darkened background. The absorption spectra of atoms are normally much simpler than the emission spectra because the transitions must start from the ground state, and only those states are excited which can be reached by a one-step transition from the ground state. Thus, comparison of absorption and emission spectra gives information about the selection rules which determine the allowed transitions, and the fact that the atom absorbs selectively from a continuous spectrum is further evidence that radiation is absorbed in discrete quanta.

Absorption spectra are particularly important in the study of the composition of stars. The very hot central region of a star radiates a continuous spectrum but certain wavelengths are absorbed by the cooler outer regions, and so by identifying these absorption lines it is possible to deduce what

*Table 5.1.* Classification of stellar spectra

| Spectral class | Temperature range (K) | Main characteristics of the spectrum |
|---|---|---|
| 0 | 30 000–50 000 | Lines from ionized helium |
| B | 10 000–30 000 | Lines from neutral helium |
| A | 7 500–10 000 | Very strong hydrogen lines |
| F | 6 000– 7 500 | Lines from ionized calcium and from many metals |
| G | 4 500– 6 000 | Strong lines from ionized calcium, ionized and neutral iron. |
| K | 3 500– 4 500 | Many lines from neutral metals |
| M | 2 000– 3 500 | Band spectra of molecules are seen, especially of titanium oxide |

elements are present in the star. There are difficulties in this analysis, however, as the atoms in the star may be highly ionized and the radiation reaching the Earth is Doppler-shifted owing to the motion of the star. Stars are classified according to their surface temperature with the hottest stars showing strong lines corresponding to hydrogen and helium and the cooler stars showing lines corresponding to much heavier elements. The classification is given in Table 5.1 and examples of stellar spectra are shown in Fig. 5.5.

## SPECTRA AND ENERGY LEVELS OF THE ALKALI ATOMS

As an example of a group of atomic systems somewhat more complicated than the hydrogen-like systems, we consider the alkali atoms lithium, sodium, potassium, rubidium. . . . These atoms have similar chemical properties and readily give up one electron to form positive ions. The ionization energy of the sodium atom is 5.1 eV, whereas the energy required to remove a second electron after the first has been removed is 47.3 eV, and the same effect is observed for the other alkali atoms. This suggests that the spectral lines observed in the infrared, visible and near ultraviolet regions of the spectrum with photon energies of a few eV are due to transitions of the least-bound electron. This electron is called the *optical* or *valence electron*.

Examination of the emission and absorption spectra of the alkali metals shows that there is a marked similarity in the appearance of the spectra of these atoms. Measurement of the spectral wavelengths and the critical potentials shows that the ground state of these atoms is not $n = 1$ as in

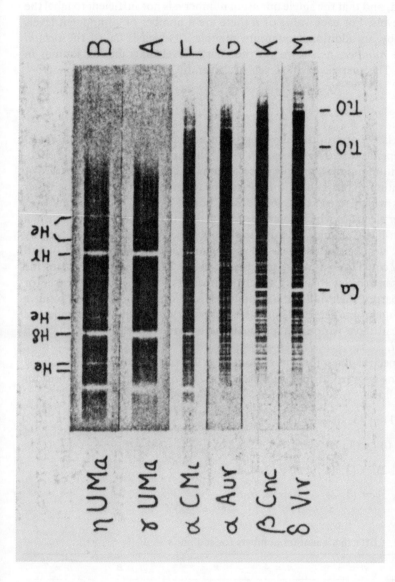

*Figure 5.5.* Examples of stellar spectra for stars at different internal temperatures. The stellar classification is given on the right and the name of the star on the left. (From J. C. Brandt, *The Physics and Astronomy of the Sun and Stars*. McGraw-Hill, 1966. Yerkes Observatory photograph.)

hydrogen, but is $n = 2$ for lithium, $n = 3$ for sodium, $n = 4$ for potassium, and so on, and that the single quantum number $n$ is not sufficient to label the energy levels. For each value of $n$ there are a number of component levels labelled by an additional quantum number $l$, which is called the *orbital quantum number*, as shown on the energy level diagram for sodium in Fig. 5.2. The energies of the levels are given by

$$W(n, l) = -\frac{hcR}{(n - \Delta)^2},\qquad(5.1)$$

where $R$ is the appropriate Rydberg constant and $\Delta$ is the *quantum defect* which depends on $l$. The values of $n - \Delta$, which can be regarded as the effective quantum number, are given for sodium in Table 5.2. The magnitude of the quantum defect can be interpreted in terms of the effect of the other electrons which screen the most loosely bound electron from the nucleus. The outer electrons which do not penetrate near to the nucleus will move in a hydrogen-like Coulomb field due to the nucleus of charge $+ Ze$ screened by the inner electrons of charge $- (Z - 1)e$, and we therefore expect $n - \Delta$ to be nearly equal to $n$; it can be seen from Table 5.2 that this is the case for the electrons in orbits with large $l$. Electrons in orbits with small $l$ experience, for at least part of their motion, a stronger effect from the nucleus and so have a larger quantum defect. Thus the electron orbits cannot all be spherical but instead have various shapes which are characterized by the quantum number $l$.

The spectral lines for these atoms can be sorted into series involving transitions between different values of $(n, l)$. For sodium, these transitions are as follows:

(i)  $(n, 1) \rightarrow (3, 0)$      $n = 3, 4, 5, \ldots$

(ii)  $(n, 0) \rightarrow (3, 1)$      $n = 4, 5, 6, \ldots$

(iii)  $(n, 2) \rightarrow (3, 1)$      $n = 3, 4, 5, \ldots$

(iv)  $(n, 3) \rightarrow (3, 2)$      $n = 4, 5, 6 \ldots$

*Table 5.2.*   Effective quantum numbers for sodium $n - \Delta$

|          | $n = 3$ | $n = 4$ | $n = 5$ | $n = 6$ | $n = 7$ |
|----------|---------|---------|---------|---------|---------|
| $l = 0$  | 1.63    | 2.64    | 3.65    | 4.65    | 5.65    |
| $l = 1$  | 2.12    | 3.14    | 4.14    | 5.14    | 6.14    |
| $l = 2$  | 2.99    | 3.99    | 4.99    | 5.99    | 6.99    |
| $l = 3$  | —       | 4.00    | 5.00    | 6.00    | 7.00    |

It can be seen that the $l$ quantum number changes by one unit in every allowed transition. This is an example of a *selection rule* and is written as

$$\Delta l = \pm 1.$$

The series (*ii*) cannot start at $n = 3$ because the level (3, 0) is below the level (3, 1). It is found experimentally that there are no levels with the quantum numbers ($n$, $n$) so that the series (*iv*) must also start at $n = 4$. The first line of each series is indicated by an unbroken arrow in Fig. 5.2 and the second line is indicated by a dashed arrow.

## ELECTRON SPIN

The transition between the (3, 1) and (3, 0) levels of sodium gives the intense radiation in the visible region whose yellow colour is taken as the signature of

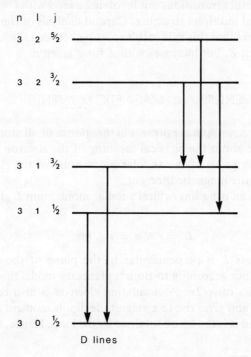

D lines

*Figure 5.6.* Doublet structure of the levels in the sodium atom due to spin–orbit splitting (not to scale).

the sodium atom. In fact, careful examination with a spectrometer shows that this radiation consists not of one line but of two lines with wavelengths of 589.0 nm and 589.6 nm, respectively. These are the so-called D-lines of sodium. All the lines in the same series and in the other series show this multiplet structure on close examination, and the number of lines observed can be explained if all the energy levels of sodium are split into two with the exception of the levels with $l = 0$. This splitting indicates the presence of yet another quantum number.

According to the interpretation introduced by Goudsmit and Uhlenbeck (1925), all electrons have a *spin quantum number* denoted by $s$, which takes the value of $\frac{1}{2}$ and adds in a rather special way to the quantum number $l$. The resultant of this addition is denoted by $j$ and has the values $l + \frac{1}{2}$ and $l - \frac{1}{2}$ with the restriction that $j$ must be positive, and electrons with the same values of $n$ and $l$ but different values of $j$ have different energies, as shown in Fig. 5.6. This theory now clearly predicts that the sodium D-lines form a doublet and the other transitions not involving a level with $l = 0$ must have a more complicated multiplet structure. Careful analysis of the spectra of the alkali atoms shows that this spin–orbit splitting between the levels decreases with $n$ for a given $Z$, but increases with $Z$ for a given $n$.

## ANGULAR MOMENTUM AND MAGNETIC MOMENTS

Effects due to electron spin are present in the spectra of all atoms. In order to understand more about the physical meaning of the electron spin, we must examine the description of the angular momentum of the electron and its connection with the magnetic moment.

An electron in an orbit has orbital angular momentum $L$ given classically by

$$L = r \wedge p = r \wedge mv.$$

For circular orbits $L$ is perpendicular to the plane of the orbit and has magnitude $mrv$, but according to Bohr's planetary model this magnitude is quantized in units of $h/2\pi$. A circulating electron is also equivalent to a current in a loop and gives rise to a magnitude dipole moment $\mu = iA$, where $i$ is the current and $A$ is the area contained by the loop. Hence, for an electron in a circular orbit

$$i = e/t, \quad t = 2\pi r/v$$

$$\therefore \; \mu_L = -e \, \frac{v}{2\pi r} \, \pi r^2 = -\frac{evr}{2} = -\frac{e}{2m} \, L.$$

The general vector relation between $\mu$ and $L$ is

$$\mu_L = -\frac{e}{2m} L$$

which is more usually written in the form

$$\mu_L = -g_l \mu_B \frac{2\pi}{h} L \qquad (5.2)$$

where $\mu_B = eh/4\pi m$ is a constant known as the *Bohr magneton* and $g_l$ is a constant known as the *g-factor*, which in this case is unity. The negative sign in these equations arises because the electron is negatively charged and implies that the magnetic moment is anti-parallel to the angular momentum, whereas for a positively charged particle in a circular orbit the magnetic moment would be in the same direction as the angular momentum.

We describe below two important experiments which were designed to measure the magnetic properties of atoms. From the Zeeman experiment, it is found that the magnitude of the orbital angular momentum $L$ for a single electron is $\sqrt{l(l + 1)}(h/2\pi)$, and not $l(h/2\pi)$ as might be expected from the planetary model, and that it is necessary to introduce a spin angular momentum $S$ and an associated magnetic moment,

$$\mu_s = -g_s \mu_B \frac{2\pi}{h} S. \qquad (5.3)$$

The magnitude of the spin angular momentum is $\sqrt{s(s + 1)}(h/2\pi)$, where $s$ is the spin quantum number introduced in the preceeding section. The value of $s$ is always $\frac{1}{2}$ for the electron, so that the spin magnetic moment is fixed and does not depend on the orbit. Since the spin magnetic moment does not arise from the motion of the electron and has no classical analogy, it is called an *intrinsic* magnetic moment. The total magnetic moment of the electron is given by

$$\mu = -\mu_B \frac{2\pi}{h} (g_l L + g_s S),$$

and the total angular momentum is

$$J = L + S,$$

which has the magnitude $\sqrt{j(j + 1)}(h/2\pi)$, where $j$ is the *quantum number of the total angular momentum*.

## STERN–GERLACH EXPERIMENT

If a magnet is placed in a non-uniform field the forces due to the field acting
on the north pole and on the south pole are not the same so that the magnet
experiences a net force which causes it to move bodily. (In a uniform field,
there would be no net translational force.) In the non-uniform field shown in
Fig. 5.7 there is a force in the $z$-direction which is given by

$$F_z = \frac{\partial B_z}{\partial z} \mu_z, \qquad (5.5)$$

where $\mu_z$ is the component of the magnetic moment in the $z$-direction.
Classically, $\mu_z = \mu \cos \alpha$ where $\alpha$ is the angle between the $z$-axis and the
direction of $\mu$, and so $\mu_z$ may take any value between $+\mu$ and $-\mu$. Hence,
classically, the force $F_z$ may take all values between $+ \partial B_z/\partial z\, \mu$ and
$- \partial B_z/\partial z\, \mu$, and since the deflection of the magnet depends on the force we
would expect to see a continuous range of deflections. The same argument
applies if we now consider a beam of atoms passing through the magnetic
field, as shown in Fig. 5.7. In fact, it is observed that the deflected beam is
always split into discrete components. This implies that the force $F_z$ takes
only discrete values, and it follows from eqn (5.5) that the orientation of the
magnetic moment with respect to the field is quantized. Hence, from eqn

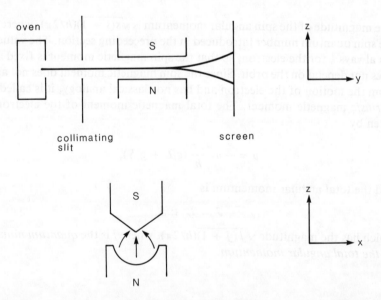

*Figure 5.7.*   A version of the apparatus used in the Stern–Gerlach experiment.

(5.2), the orientation of the angular momentum is quantized, i.e. not only the magnitude but also the direction of the angular momentum is quantized.

This experiment was first performed by Stern and Gerlach in 1922 using silver atoms, but has since been repeated with many atomic beams. We consider the case when the beam is composed of lithium atoms in the ground state, which have the quantum numbers $n = 2, l = 0, s = \frac{1}{2}$. It is found that the beam is split into two components and the same result is observed for all atoms which have a single optical electron with $l = 0$. In this case, it can be seen from eqn (5.2) that the orbital magnetic moment is zero and the occurrence of the deflection is evidence for the existence of the spin magnetic moment defined in eqn (5.3). If we denote the $z$-component of the spin angular momentum by $m_s(h/2\pi)$, the $z$-component of the magnetic moment is

$$\mu_{sz} = -g_s\mu_B m_s. \tag{5.6}$$

Since $s = \frac{1}{2}$, we would expect that $|m_s| \leqslant \frac{1}{2}$, and the fact that the beam is split into two components suggests that

$$m_s = +\tfrac{1}{2} \quad \text{or} \quad -\tfrac{1}{2},$$

i.e. that there are $(2s + 1)$ components. The magnitude of the splitting then gives the magnitude of the $g$-factor as $g_s = 2$.

If the atom is in a state with $l \neq 0$, the total magnetic moment is given by

$$\mu = -\mu_B \frac{2\pi}{h}(L + 2S), \tag{5.7}$$

and the observed splitting of the beam is consistent with a $z$-component of the orbital angular momentum given by $m_l(h/2\pi)$ with

$$m_l = -l, -l + 1, \ldots -1, 0, 1 \ldots +l,$$

Thus, $m_l$ can take $2l + 1$ integer values between $-l$ and $+l$. Using eqn (5.4) the magnetic moment given by eqn (5.7) can be rewritten as

$$\mu = -\mu_B \frac{2\pi}{h}(J + S),$$

from which it can be seen that the total magnetic moment $\mu$ is not, in general, anti-parallel to the total angular momentum $J$.

ZEEMAN EFFECT

In order to understand the Zeeman effect it is easiest to think of the electron in a circular orbit. When the circulating electron is placed in a constant magnetic field, the orbit is subject to a torque which causes the angular momentum vector to precess about the field direction. (This is the same effect as the precession of a spinning top.) The change in the energy of the system is

$$\Delta E = -\mu \cdot B$$

and, with the magnetic moment given by eqn (5.2), this change in energy due to the effect of a uniform magnetic field is given by

$$\Delta E = g_l \mu_{\mathrm{B}} \frac{2\pi}{h} L \cdot B,$$

but if we choose the z-axis along the direction of the magnetic field we have

$$\frac{2\pi}{h} L \cdot B = m_l B$$

and

$$\Delta E = g_l \mu_B m_l B.$$

It can be shown, but will not be proved here, that when the spin of the electron is included so that $\mu$ is given by eqn (5.7), the change in the energy of the electron in the presence of the field is

$$\Delta E = g \mu_{\mathrm{B}} m_j B, \tag{5.8}$$

where $g$ is a new $g$-factor which depends on the values of $l$, $s$ and $j$. For $l = 0$, $g = g_s = 2$. The quantum number $m_j$ is the z-component of the total angular momentum and has $2j + 1$ values from $-j$ to $+j$. Examination of the optical spectra emitted by atoms in a fairly weak magnetic field shows that the lines have a structure which is consistent with the splitting of the energy levels predicted by eqn (5.8). This effect was first observed by Zeeman in 1896. The structure of the sodium D-lines is shown in Fig. 5.8.

Measurements of the splitting observed in the Zeeman effect yield values of the $g$-factor, and from these values it can be deduced that the magnitude of the total angular momentum $J$ is indeed given by $\sqrt{j(j + 1)}(h/2\pi)$, and similarly that the magnitudes of the orbital and spin angular momenta are $\sqrt{l(l + 1)}(h/2\pi)$ and $\sqrt{s(s + 1)}(h/2\pi)$, respectively.

## FURTHER REMARKS ABOUT QUANTUM NUMBERS

It should be noted that the spatial quantization of angular momentum only becomes apparent when an external field (magnetic or electric) is applied to

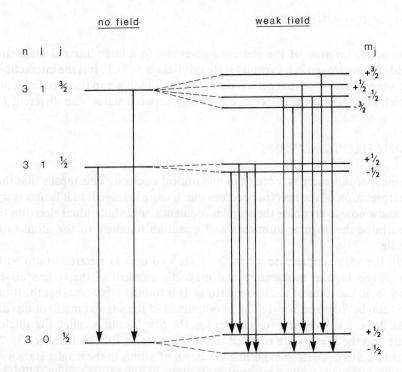

*Figure 5.8.* The splitting of the levels in the sodium atom due to the effect of a magnetic field, and the resulting structure of the D-lines observed in the Zeeman effect.

impose a direction of quantization on the system. For this reason, the quantum numbers $m_l$, $m_s$ and $m_j$ are sometimes called magnetic quantum numbers.

In the absence of an external field, electrons with the same $n$, $l$, $j$ and $m_j$ have the same energy. In the absence of both an external field and the spin–orbit splitting, electrons with the same $n$ and $l$ but different $j$ and $m_j$ have the same energy, and electrons with the same $n$ but different and large $l$ have almost the same energy (see Fig. 5.2). For this reason we regard $n$ as the quantum number which primarily determines the energy of the electron. This interpretation of $n$ is in accordance with Bohr's planetary model, but we have clearly departed from the planetary model in that $n$ is no longer connected with the orbital angular momentum. The latter quantity is now associated with a different quantum number, namely $l$.

## SPIN–ORBIT INTERACTION

The orbital motion of the electrons gives rise to a large internal magnetic field. For hydrogen in its ground state this field is ~ 50 T. It is the interaction between the intrinsic magnetic moment due to the spin and this internal field which gives rise to the difference in energy between states with different $j$.

## MANY-ELECTRON ATOMS

In most atoms there is more than one optical electron. This means that the interpretation of the spectra becomes much more involved, as it is necessary to know how to combine the angular momenta of the individual electrons to determine the angular momenta and quantum numbers of the atom as a whole.

In the very simple case when there are two optical electrons both with $l = 0$, the angular momentum and magnetic moment of the system arises only from the spins of the two electrons. It is found in this case that the total spin can be $S = 0$ or $S = 1$. The $z$-component of the angular momentum can take $2S + 1$ values, and for this reason the $S = 0$ state is called the singlet state and the $S = 1$ state is called the triplet state. It follows from eqn (5.7) that in a Stern–Gerlach experiment a beam of atoms in the singlet state will not be deflected, while a beam of atoms in the triplet state will be split into three components.

## X-RAY SPECTRA

The production of the continuous spectrum of X-ray radiation was described in Chapter 2. It was noted that a discrete line spectrum can also be produced, and it is this line spectrum which is relevant to the present examination of the evidence for quantization.

If one of the bombarding electrons in an X-ray tube knocks out one of the tightly bound electrons from an atom in the target, the atom is left with a vacancy in one of its low-lying levels. An electron from a higher level can then fall to the lower, more tightly bound, level giving out energy in the form of a quantum of radiation whose energy is determined by the difference between the energies of the two levels. The targets used in X-ray tubes are normally composed of elements of high $Z$ such as tungsten ($Z = 74$) or molybdenum ($Z = 42$) so that the low-lying levels have binding energies of the order of keV, and hence the emitted radiation is in the keV region. An energy level diagram for X-ray transitions is shown in Fig. 5.9. Transitions

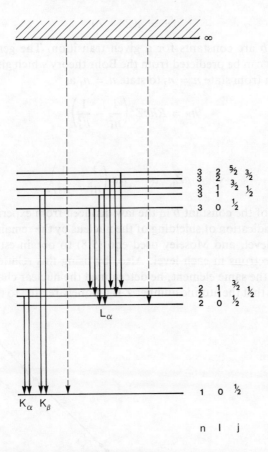

*Figure 5.9.* Transitions between atomic energy levels leading to emission of characteristic X-ray lines.

which end on the $n = 1$ level form of K-series, those which end on the $n = 2$ level form the L-series, and so on. The short-wavelength limit for each series is called the *absorption edge*. The X-ray lines show multiplet structure due to the splitting of levels with the same $n$ but different $l$ and $j$.

## MOSELEY'S LAW

Moseley (1913) studied the X-ray line spectra of many elements and found that for a given transition the frequencies of the lines emitted by different elements obeyed the relation

$$\sqrt{\nu} = a(Z - b), \qquad (5.8)$$

where $a$ and $b$ are constants for a given transition. The general form of Moseley's law can be predicted from the Bohr theory which gives the energy of a transition from state $n = n_2$ to state $n = n_1$ as

$$h\nu = RhcZ^2\left(\frac{1}{n_1^2} - \frac{1}{n_2^2}\right)$$

or

$$\sqrt{\nu} = aZ, \quad a^2 = Rc\left(\frac{1}{n_1^2} - \frac{1}{n_2^2}\right).$$

The presence of the constant $b$ in the law deduced from experiment is interpreted as an indication of shielding of the nucleus by the remaining electrons in the lower level, and Moseley used eqn (5.8) to obtain estimates of the number of electrons in each level. Also, by using this relation for several transitions in the same element, he determined the nuclear charge $+ Ze$ and demonstrated that the atomic number $Z$ is uniquely related to the position of

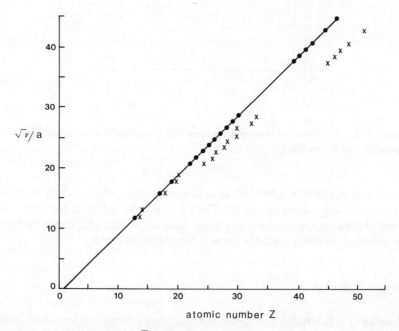

*Figure 5.10.* Behaviour of $\sqrt{\nu}/a$ for the transition $n = 2$ to $n = 1$ plotted against atomic number $Z$ (dots) and against atomic weight (crosses) on a different scale.

the element in the periodic table and does not necessarily follow the atomic weight. For example, Fig. 5.10 shows a plot of the quantity $\sqrt{\nu}/a$, where $a^2 = \frac{3}{4}Rc$ for the transition from state $n_2 = 2$ to state $n_1 = 1$, against atomic number $Z$ and against atomic weight. The intercept on the $x$-axis gives $b \approx 1$ in this case. This indication of the significance of $Z$ was of great importance at the time, and is now quite understandable from our knowledge of the existence of isotopes.

## DEFICIENCIES IN THE BOHR THEORY

In Chapter 4, the Bohr theory was applied to the hydrogen atom, and it proved possible to calculate with high accuracy the location of the energy levels and the frequency of the radiation associated with transitions between the levels. However, the theory used in Chapter 4 has certain obvious deficiencies which may be stated as follows:

(*i*)  It cannot be used to make any calculations about the transitions, such as the rate at which they occur or the selection rules which apply to them.

(*ii*)  The planetary model introduces only one quantum number, whereas there is experimental evidence for the existence of additional numbers.

(*iii*)  The theory applies to one-electron atoms, and is not easily extended to describe more complicated atoms.

As an illustration of the difficulties arising from the last two points, we use the Bohr theory to calculate the ionization energy of the next two elements, helium with $Z = 2$ and lithium with $Z = 3$. According to the Bohr theory the energy of an electron with quantum number $n$ is $W_n = -hcRZ^2/n^2$, so that the ionization energy required to free the electron from its atom is just $|W_n|$. As shown in Chapter 4, the Rydberg constant is not quite constant for different atoms but we can neglect the small variation for the purpose of this

*Table 5.3.*  Ionization energies for hydrogen, helium and lithium

| | $Z$ | Observed ionization energy (eV) | | Calculated ionization energy (eV) | |
|---|---|---|---|---|---|
| | | First | Second | $n = 1$ | $n = 2$ |
| H | 1 | 13.6 | — | 13.6 | — |
| He | 2 | 24.5 | 54.1 | 54.4 | — |
| Li | 3 | 5.4 | | 122.4 | 30.8 |

example. Inserting the appropriate values of $Z$ we obtain the calculated ionization energies given in Table 5.3. The experimental values listed are the energies required to remove one electron from the neutral atom (first ionization energy) and the energies required to remove a second electron after the first has been removed (second ionization energy). It can be seen from the table that if we assume that both electrons in helium are in the $n = 1$ state, the Bohr theory is accurate for the singly ionized helium ion $He^+$ which has only one electron but fails for the helium atom which has two electrons. This result may be interpreted as an indication of the importance of the repulsive interaction between the two electrons which tends to reduce the effective binding of each electron to the nucleus. From the Bohr theory, we might expect that the first ionization energy for lithium would be greater than that for helium owing to the increase in $Z$, but instead the observed value drops substantially below the value for hydrogen. This suggests that the least-bound electron in the lithium atom is not in the level corresponding to $n = 1$ but is in the $n = 2$ or some higher level. The information obtained from a study of the spectrum indicates that this least-bound electron in lithium is indeed in the $n = 2$ level, but gave no indication why this is the case. Hence, in order to make any calculations at all for many-electron atoms we need some additional rule or principle which will determine how many electrons go into each energy level.

It should be possible to estimate the repulsive force between the two electrons in the helium atom by averaging the repulsion when the electrons are in various positions in the two orbits, but in order to do this we need to know the orientation of one of the orbits relative to the other, and the simple planetary model does not provide this information because it is a one-dimensional model. The planetary model can be extended to allow for more complicated orbits and this was first done by Sommerfeld, who introduced the idea of elliptical orbits, but the theory remains an uneasy combination of quantum and classical ideas. We therefore leave the Bohr theory at this point and, while continuing to use the ideas of quantization and allowed energy states, we seek to understand how a more complete and rigorous quantum theory can be formulated.

## THE EXCLUSION PRINCIPLE

The rule which determines the maximum number of electrons in a given level was deduced by Pauli (1925) from a study of experimental data on the spectra of helium and similar atoms. The rule states that no two electrons can have the same set of quantum numbers.

The quantum numbers which describe an individual electron in an atom were introduced in connection with the experimental study of spectra and ionization energies, and may be summarized as follows:

$n$
- Principle quantum number which takes integer values from 1 to $\infty$.
- Determines the size of the electron orbit.
- Arises from quantization of energy.

$l$
- Orbital quantum number which takes integer values from 0 to $n - 1$.
- Determines the shape of the electron orbit.
- Arises from the quantization of the magnitude of the orbital angular momentum.

$m_l$
- Magnetic quantum number which takes integer values from $-l$ to $+l$.
- Determines the orientation of the orbit in space.
- Arises from the quantization of the direction of the orbital angular momentum.

$s$
- Electron spin quantum number which takes the value $\frac{1}{2}$ for all electrons.
- Arises from the intrinsic spin angular momentum of the electron.

$m_s$
- Spin projection quantum number which takes values of $\pm \frac{1}{2}$ only.
- Determines the spin orientation, *up* or *down*.
- Arises from the quantization of the direction of the spin angular momentum.

Thus, the four quantum numbers $n$, $l$, $m_l$ and $m_s$ (or $n$, $l$, $j$ and $m$) can be regarded as *labels* which specify completely the state of an electron in an atom, and hence determine its energy and angular momentum. When an electron makes a transition from one state to another, some or all of these quantum numbers and the associated properties will change. On the other hand, such properties as charge, rest mass and spin are intrinsic properties of the electron and do not change.

For a normal atom in the absence of any external fields the exclusion principle can be used to determine the maximum number of electrons which can go into a given level. For example, the level with $n = 2$ and $l = 0$ has $m_l = 0$ so that there is a maximum of two electrons, one with $m_s = +\frac{1}{2}$ and one with $m_s = -\frac{1}{2}$, which can go into this level. The level with $n = 2$ and $l = 1$ has $m_l = 0, \pm 1$ so that there is a maximum of six possible combinations of $m_l$ and $m_s$ in this case. As a result of the spin–orbit interaction this level is split into two components, one with $j = \frac{3}{2}$ and $(2j + 1) = 4$ values of $m_j$ and one with $j = \frac{1}{2}$ and $(2j + 1) = 2$ values of $m_j$, and there are still six possible combinations of the quantum numbers $j$ and $m_j$. In the discussion that follows we neglect the spin–orbit interaction and label the electrons with $n$, $l$, $m_l$ and $m_s$.

## ELECTRON SHELLS AND THE PERIODIC TABLE

Electrons that have the same quantum number $n$ are said to belong to the same *shell*. These shells are denoted by letters such that the K-shell has $n = 1$, the L-shell has $n = 2$, and so on, in accordance with the notation for the series of X-ray transitions. For a given value of $n$, electrons having the same quantum number $l$ are said to belong to the same *subshell*, and the subshells are also denoted by letters, using the notation given in Table 5.4. The maximum number of electrons in a subshell is given by $2(2l + 1)$ since there are $2l + 1$ values of $m_l$ and two values of $m_s$. Hence the maximum number of electrons in a shell is given by

$$N(n) = \sum_{l=0}^{n-1} 2(2l + 1) = 2n^2. \tag{5.9}$$

It is now possible to deduce the electronic configuration of atoms by filling up each shell in turn, starting with the most strongly bound. These predicted configurations and the observed ionization energies are given in Table 5.5. It can be seen that when a shell or subshell is closed the ionization energy rises, but when there is a single electron in a shell the ionization energy is very low and the electron is readily detached. It is also evident that the ordering of elements is naturally associated with the atomic number $Z$ and not the atomic weight. Comparison with the Periodic Table shown in Fig. 5.11 shows that similarity in electronic configuration corresponds to similarity in chemical properties. For nuclei with $Z > 18$ the ordering of the levels becomes irregular and must be deduced from detailed study of the atomic spectra, but once the electronic configuration has been established many of the chemical and physical properties of the element can be understood.

It is now possible to examine the effect of the exclusion principle on the possible transitions between levels, and to interpret the experimental observations described earlier in this chapter. As an example we consider transitions in the sodium atom. We see from Table 5.5 that in the normal unexcited atom of sodium, the K- and L-shells are filled and there is one electron in the 3s level of the M-shell. This explains why the 3s level is the ground state of sodium, as shown in Fig. 5.2, and not the 1s level. The single electron in the M-shell can be excited from the 3s level to higher levels in the same shell or into a higher shell. This excitation cannot involve energies

*Table 5.4.* Notation for subshells

| Orbital angular momentum $l$ | 0 | 1 | 2 | 3 | 4 | ... |
|---|---|---|---|---|---|---|
| Notation for subshell | s | p | d | f | g | ... |
| Number of electrons in subshell | 2 | 6 | 10 | 14 | 18 | ... |

*Table 5.5.* Electronic configurations and ionization energies

| Z | Element | K-shell $n=1$ $l=0$ 1s | L-shell $n=2$ $l=0$ 2s | $l=1$ 2p | M-shell $n=3$ $l=0$ 3s | $l=1$ 3p | $l=2$ 3d | N-shell $n=4$ $l=0$ 4s | Ionization energy (eV) |
|---|---|---|---|---|---|---|---|---|---|
| 1 | H | 1 | | | | | | | 13.6 |
| 2 | He | 2 | | | | | | | 24.5 |
| 3 | Li | 2 | 1 | | | | | | 5.4 |
| 4 | Be | 2 | 2 | | | | | | 9.3 |
| 5 | B | 2 | 2 | 1 | | | | | 8.3 |
| 6 | C | 2 | 2 | 2 | | | | | 11.2 |
| 7 | N | 2 | 2 | 3 | | | | | 14.5 |
| 8 | O | 2 | 2 | 4 | | | | | 13.6 |
| 9 | F | 2 | 2 | 5 | | | | | 17.3 |
| 10 | Ne | 2 | 2 | 6 | | | | | 21.5 |
| 11 | Na | 2 | 2 | 6 | 1 | | | | 5.1 |
| 12 | Mg | 2 | 2 | 6 | 2 | | | | 7.6 |
| 13 | Al | 2 | 2 | 6 | 2 | 1 | | | 6.0 |
| 14 | Si | 2 | 2 | 6 | 2 | 2 | | | 8.1 |
| 15 | P | 2 | 2 | 6 | 2 | 3 | | | 10.9 |
| 16 | S | 2 | 2 | 6 | 2 | 4 | | | 10.3 |
| 17 | Cl | 2 | 2 | 6 | 2 | 5 | | | 12.6 |
| 18 | Ar | 2 | 2 | 6 | 2 | 6 | | | 15.7 |
| 19 | K | 2 | 2 | 6 | 2 | 6 | | 1 | 4.3 |
| 20 | Ca | 2 | 2 | 6 | 2 | 6 | | 2 | 6.1 |
| 21 | Sc | 2 | 2 | 6 | 2 | 6 | 1 | 2 | 6.5 |
| 22 | Ti | 2 | 2 | 6 | 2 | 6 | 2 | 2 | 6.8 |
| 23 | V | 2 | 2 | 6 | 2 | 6 | 3 | 2 | 6.7 |
| 24 | Cr | 2 | 2 | 6 | 2 | 6 | 5 | 1 | 6.8 |
| 25 | Mn | 2 | 2 | 6 | 2 | 6 | 5 | 2 | 7.4 |
| 26 | Fe | 2 | 2 | 6 | 2 | 6 | 6 | 2 | 7.9 |
| 27 | Co | 2 | 2 | 6 | 2 | 6 | 7 | 2 | 7.9 |
| 28 | Ni | 2 | 2 | 6 | 2 | 6 | 8 | 2 | 7.6 |
| 29 | Cu | 2 | 2 | 6 | 2 | 6 | 10 | 1 | 7.7 |
| 30 | Zn | 2 | 2 | 6 | 2 | 6 | 10 | 2 | 9.4 |

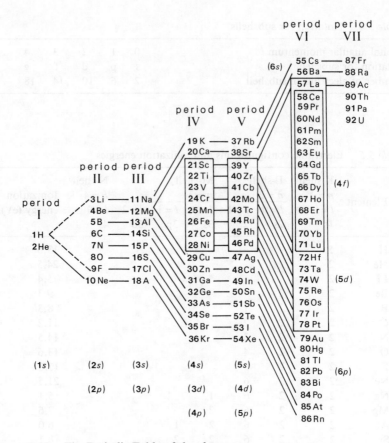

*Figure 5.11.*   The Periodic Table of the elements.

greater than 5.1 eV since these energies cause removal of the electron from the atom. De-excitation from these higher levels back to the 3s level gives rise to emission of radiation with photon energies up to 5.1 eV, i.e. radiation in the visible and infrared regions. Excitation of the electrons in the K- and L-shells into the M- or higher shells requires energies in the keV region and de-excitation gives rise to the emission of X-ray radiation. When the atom is in its ground state, excitation of an electron from the K- to L-shell is forbidden by the exclusion principle as the L-shell is already full. Thus, the effect of the exclusion principle is such that we can think of the full or *closed* shells as a relatively stable and inert core while the outer unfilled shell contains the active electrons whose behaviour determines most of the chemical and physical properties. As a contrast, it is interesting to speculate about the nature of atomic properties if the exclusion principle did not apply. In this case, an atom in its ground state would have every electron in the 1s level so that it would be reduced to the smallest possible size and, as all properties would be determined by the number of electrons, there would be no periodicity in the chemical properties or in the spectra.

## PROBLEMS

**5.1.** Using the information given in Fig. 5.1, calculate the short-wavelength limit of the Balmer series for hydrogen and check your result using Table 4.1. Calculate also the longest wavelength line in the series terminating on state $n = 3$.

*Solution.* $\lambda_{min} = 365$ nm, $\lambda_{max} = 1885$ nm

**5.2.** Using the information given in Fig. 5.1, find the values of the following critical potentials for the hydrogen atom:

(*i*)  The ionization potential.

(*ii*)  The resonance potential.

(*iii*)  The excitation potential for the state with $n = 4$.

*Solution.*  (*i*)  13.6 eV
(*ii*)  10.2 eV
(*iii*)  12.8 eV

**5.3.** The first, second and third excited states of a hypothetical atom lie at 2.1, 3.3 and 3.9 eV, respectively, above the ground state. These atoms in the form of a gas are bombarded with electrons which have been accelerated through a voltage of 3.6 V. Explain what is observed and calculate the possible final values of kinetic energy of the electrons. What will be observed if the accelerating voltage is reduced to 1.8 V?

*Solution.*  Possible kinetic energy = 3.6 eV, 1.5 eV, 0.3 eV

**5.4.** Atoms of the type described in the previous question are bombarded with photons of selected radiation. Explain what is observed if the photon energy is:

(*i*)  3.6 eV,

(*ii*)  2.1 eV.

**5.5.** Using the values given in Table 5.2, find the magnitudes of the quantum defect for sodium and verify that they are independent of the quantum number $n$.

*Solution.* $l = 0, \Delta = 1.35; l = 1, \Delta = 0.86; l = 2, \Delta = 0.01$

**\*5.6.** The longest wavelength lines in the series $(n, 1) \rightarrow (4, 0)$ in potassium have wavelengths of 769.90, 766.49, 404.72 and 404.41 nm, respectively. Construct the relevant part of the energy level diagram and calculate the splitting between the levels with the same values of $n$ and $l$ but different $j$.

*Solution.* The spectrum shows the characteristic doublet structure due to the electron spin. The first two lines arise from transitions from the pair of levels with $n = 4, l = 1, j = \frac{3}{2}, \frac{1}{2}$ to the ground state, $n = 4, l = 0, j = \frac{1}{2}$, while the other pair of lines arises from transitions from the higher levels with $n = 5, l = 1, j = \frac{3}{2}, \frac{1}{2}$ to the ground state. Thus there are five levels to be drawn on the energy level diagram, two pairs of levels with $l = 1$ and a single level (the ground state) with $l = 0$.

The separation between the levels can be calculated using the formula

$$\Delta E = hc\left(\frac{1}{\lambda_1} - \frac{1}{\lambda_2}\right) = hc\left(\frac{\lambda_2 - \lambda_1}{\lambda_1 \lambda_2}\right).$$

For example, the separation between the two levels with $n = 4, l = 1$ is given by

$$\Delta E = 6.625 \times 10^{-34} \times 3 \times 10^8 \times \frac{34.1 \times 10^{-10}}{7699.0 \times 7664.9 \times 10^{-20}} \text{ J}$$

$$= \frac{6.625 \times 3 \times 34.1 \times 10^{-22}}{7.70 \times 7.66 \times 1.6 \times 10^{-19}} \text{ eV}$$

$$= 7.2 \times 10^{-3} \text{ eV}.$$

**\*5.7.** By examining Fig. 5.9 show that the X-ray transitions satisfy the selection rule $\Delta l = \pm 1$, and find the selection for the quantum number j.

**5.8.** The K absorption edge in tungsten is $\lambda_\infty = 0.0178$ nm and the average wavelengths of some lines in the K-series are $\lambda_\alpha = 0.0210$ nm, $\lambda_\beta = 0.0184$ nm, $\lambda_\gamma = 0.0179$ nm. If a tungsten target is bombarded with electrons of 100 keV calculate the maximum kinetic energies of the electrons emitted from the levels with $n = 1, n = 2$ and $n = 3$.

*Solution.* The wavelengths of the K-series lines are given by

$$\frac{hc}{\lambda_\alpha} = W_2 - W_1, \qquad \frac{hc}{\lambda_\beta} = W_3 - W_1,$$

$$\frac{hc}{\lambda_\gamma} = W_4 - W_1, \quad \frac{hc}{\lambda_\infty} = W_\infty - W_1.$$

But by definition $W_\infty = 0$. Hence

$$W_1 = -\frac{6.625 \times 10^{-34} \times 3 \times 10^8}{0.178 \times 10^{-10}} \frac{1}{1.6 \times 10^{-19}} \text{ eV}$$

$$= -69.6 \text{ keV}.$$

The energy of the next level is given by

$$W_2 = \frac{hc}{\lambda_\alpha} + W_1,$$

and similarly for $W_3$ and $W_4$. The maximum kinetic energy of the electrons emitted from each level is given by

$$T_{\max} = 100 - |W_n|.$$

This assumes that the bombarding electron is reduced to rest and the emitted electron carries away all the excess energy, but this energy may, of course, be shared between the two electrons.

**\*5.9.** Show that the excited states of the helium atom can have total spin of zero (parahelium) or unity (orthohelium), but that the ground state can have a spin of zero only. It is observed experimentally that there are no transitions between states of parahelium and orthohelium; from this information deduce a selection rule for transitions in helium and comment on its interpretation.

**5.10.** The ionization energies necessary to remove a second electron from singly ionized ions from sodium to calcium are tabulated below. Explain the difference in magnitude and variation with atomic number between these energies and the first ionization energies given in Table 5.5.

| Element | Na | Mg | Al | Si | P | S | Cl | A | K | Ca |
|---------|------|------|------|------|------|------|------|------|------|------|
| 2nd IE (eV) | 47.3 | 15.0 | 18.8 | 16.3 | 19.7 | 23.4 | 23.8 | 27.6 | 31.8 | 11.9 |

# 6 | Development of the Quantum Theory

## CRITICISM OF THE BASIS OF CLASSICAL MECHANICS

The basis of classical mechanics depends on two fundamental assumptions about the system and the relation between the observer and the system:

(*i*)   It is assumed that it is possible to determine precisely the initial state of the system at time $t_0$. The initial state is characterized by the position and velocity of each particle at time $t_0$, and the equations of motion determine the state of the system at all later times. Thus, classical mechanics is a *deterministic* theory* describing a *causal* system.

(*ii*)   The existence of an isolated system, or one influenced by known forces, is postulated. This implies the presence of an observer who can record the state of the system at time $t_0$ without disturbing the system by making the measurement.

These assumptions are in conflict with the uncertainty principle, which shows that there is a fundamental limit to the accuracy with which we can define the initial system and that the measurement of one variable introduces an uncertainty into another, and it follows that classical mechanics is not valid for atomic systems. It also follows that the quantum theory which replaces classical mechanics for the description of atomic systems must reflect the indeterminism arising from the uncertainty principle, although

---

* It is known that even classical determined systems can show apparently random behaviour which is called *chaotic*. The randomness here arises because of extreme sensitivity to the initial conditions. For example, very small changes in the assumed initial conditions can have dramatic effects on the outcome of predictions of weather patterns some time ahead. In chaotic systems there is an underlying determinism.

this does not mean that this theory can be in any way indefinite because we must be able to make definite predictions for observable quantities.*

## IMPLICATIONS OF WAVE–PARTICLE DUALITY

It is evident that the description of electrons in terms of quantum numbers is a very powerful tool, but we have as yet no rigorous theoretical basis from which to derive all the required quantum numbers. It seems necessary to depart from the classical descriptions of particle motion, and an indication of the way forward comes from recognition of the fact that the description of wave motions in classical physics contains many of the features which are required in the quantum description of electrons in atoms. The vibrating stretched string, and also the vibrating membrane of a drum, have the characteristic that they vibrate only in certain allowed modes and that the discrete set of allowed values of the wavelength is determined by a set of integers. The same is true of electromagnetic radiation in a black box or cavity. In addition, we have seen in Chapter 2 that electromagnetic radiation shows both wave and particle properties. The next step, therefore, is to examine the mathematical properties of wave motions.

## MATHEMATICAL REPRESENTATION OF WAVE MOTIONS

A wave motion may be generally defined as the propagation of a disturbance in a periodic manner from one point to another in a medium, without any bodily motion of the medium itself. Wave motions are of two types: for longitudinal waves the individual particles of the medium vibrate along the direction of propagation, whereas for transverse waves the vibration is perpendicular to the direction of propagation. In either case, the displacement of the individual particles as a result of the passage of a wave can be written as

$$\eta = a \sin\left(2\pi\nu t - \frac{2\pi}{\lambda} x\right), \tag{6.1}$$

where $\eta$ is the *displacement* measured for particular values of the position $x$ and time $t$, $a$ is the *amplitude*, which is the maximum value of the displacement, and $\nu$ is the *frequency*, which is the number of complete vibrations per

---

* In quantum systems, it is generally assumed that no underlying determinism exists, although some distinguished scientists, e.g. Einstein and D. Bohm, have disputed the view.

unit time. Two other important associated parameters are the *period* $T = 1/\nu$, which is the time taken for an individual particle to make one complete vibration, and the *wavelength* $\lambda$, which is the shortest distance between two particles whose displacements and instantaneous direction of motion are the same at a given time, i.e. the distance travelled in one period. Equation (6.1) can also be written as

$$\eta = a \sin (\omega t - kx), \tag{6.2}$$

where $\omega = 2\pi\nu$ is the *angular frequency* and $k = 2\pi/\lambda$ is the *wave number*.*

The choice of a sine function automatically ensures that the motion represented by eqns (6.1) or (6.2) is periodic, i.e. the displacement has the same value for $x$ and $x + \lambda$, and for $t$ and $t + T$. The same periodicity could be represented by a cosine function or a combination of cosine and sine functions. The distance travelled by the wave in the time $T$ is $uT$, where $u$ is the *wave velocity*, but by definition this distance is also equal to the wavelength, so that $\lambda = uT$, or

$$u = \nu\lambda = \omega/k. \tag{6.3}$$

The displacement due to two or more wave motions is found by using the *principle of superposition*, which states that the resultant displacement at any point and time is found by adding the displacements which would be produced if each wave were present alone. As an example, we add the displacements due to two waves of equal amplitude whose wave numbers and angular frequencies differ by small amounts. The resultant displacement is given by

$$\eta = \eta_1 + \eta_2$$
$$= a \sin (\omega_1 t - k_1 x) + a \sin (\omega_2 t - k_2 x)$$
$$= 2a \cos \left( \frac{\omega_1 - \omega_2}{2} t - \frac{k_1 - k_2}{2} x \right) \sin \left( \frac{\omega_1 + \omega_2}{2} t - \frac{k_1 + k_2}{2} x \right).$$

We now let $k_1 = k_2 + \Delta k$, $\omega_1 = \omega_2 + \Delta\omega$, so that

$$\eta = 2a \cos (\tfrac{1}{2}\Delta\omega t - \tfrac{1}{2}\Delta kx) \sin \{(\omega_1 - \tfrac{1}{2}\Delta\omega)t - (k_1 - \tfrac{1}{2}\Delta k)x\}.$$

Now, if $\Delta\omega$ and $\Delta k$ are small, the behaviour of the sine term is not very different from either of the original wave motions, but the cosine term varies very slowly, so that the combined behaviour is as shown in Fig. 6.1. The region of large amplitude, called a *wave packet* or *group*, travels with a low

---

* The definition of the wave number given here is the one commonly used in quantum theory, many-body theory and nuclear physics. In spectroscopy, however, it is customary to define the wave number to be $1/\lambda$ and denote it by the symbol $\bar{\nu}$.

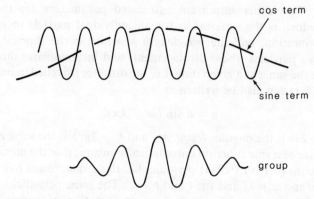

cos term

sine term

group

*Figure 6.1.* Formation of a group due to the superposition of displacements due to two wave motions.

velocity compared with the constituent wave motions which move through it. The velocity of the wave packet is determined by comparing values of $x$ and $t$ for which the argument of the cosine function is the same, i.e.

$$\tfrac{1}{2}\Delta\omega t_1 - \tfrac{1}{2}\Delta k x_1 = \tfrac{1}{2}\Delta\omega t_2 - \tfrac{1}{2}\Delta k x_2$$

$$\tfrac{1}{2}\Delta k(x_2 - x_1) = \tfrac{1}{2}\Delta\omega(t_2 - t_1),$$

and hence the velocity is given by

$$U = \frac{x_2 - x_1}{t_2 - t_1} = \frac{\Delta\omega}{\Delta k}.$$

The length of the wave packet is the distance $l = x_2 - x_1$ travelled as the cosine function through one half-cycle, i.e.

$$\tfrac{1}{2}\Delta k\, l = \pi$$

$$\therefore l = 2\pi/\Delta k.$$

In simple applications of these formulae it is assumed that the wave velocity $u$ is a constant independent of wavelength, which is correct for sound waves. Media in which the wave velocity is a function of the wavelength are called *dispersive*, and in these media there is a tendency for disturbances of different wavelengths to form groups. We are now dealing with a continuous spread of $\omega$ and $k$ so that the addition of the displacements is more complicated. It is found that the group velocity is given by

$$U = \frac{d\omega}{dk} \qquad\qquad (6.4)$$

and the length of the group is given by

$$l = \frac{2\pi}{\mathrm{d}k}. \tag{6.5}$$

Thus, a group which is small in spatial extent corresponds to a large value of $\mathrm{d}k$, i.e. to a large spread of wavelengths for the constituent waves, and vice versa.

## THE WAVE EQUATION AND ITS SOLUTION*

Many of the laws of physics can be expressed in terms of rates of change. For example, the basic equation of Newtonian mechanics is

$$\frac{\mathrm{d}}{\mathrm{d}t}(mv) = F,$$

where $F$ is the force and $mv$ is the momentum. (If $m$ is constant, the rate of change of momentum is just mass times acceleration.) This is an ordinary differential equation.

In order to find the form of the wave equation, we must seek the differential equation obeyed by the displacement $\eta$ as a function of the space variables and time. For simplicity we continue to consider only one space variable $x$. Differentiating eqn (6.2) twice with respect to $x$ we have

$$\frac{\partial^2\eta}{\partial x^2} = -k^2 a \sin(\omega k - kx) = -k^2\eta,$$

and differentiating with respect to $t$ we have

$$\frac{\partial^2\eta}{\partial t^2} = -\omega^2 a \sin(\omega k - kx) = -\omega^2\eta.$$

Combining these two results we obtain the wave equation in one dimension

$$\frac{\partial^2\eta}{\partial t^2} = \frac{\omega^2}{k^2}\frac{\partial^2\eta}{\partial x^2}. \tag{6.6}$$

This equation describes many different wave phenomena. The solution can be written in a more general form than (6.2), but the particular form

* This section can be omitted on a first reading, particularly if the reader is unfamiliar with partial differentiation. However, a knowledge of partial differentiation and partial differential equations is eventually necessary for a full understanding of classical wave theory and quantum theory.

appropriate for a given problem is determined only when we apply the initial and boundary conditions appropriate to that problem. As an example of the application of the conditions we consider the case of a stretched string of length $L$ which is fixed at each end and plucked at time $t = 0$. The mathematical statement of the restraints on the system is

$$\eta = 0 \quad \text{at } x = 0 \text{ and } x = L, \text{ for all } t.$$

The condition for $x = 0$ is satisfied for a solution of the form

$$\eta = a \sin kx \cos \omega t.$$

The remaining condition imposes the requirement that

$$0 = a \sin kL,$$

which can be satisfied only if $kL = n\pi$ where $n$ is an integer. Hence

$$k = \frac{n\pi}{L}, \quad \lambda = \frac{2L}{n}, \tag{6.7}$$

and the string vibrates only in certain *allowed modes of vibration* with *characteristic* wavelengths determined by eqn (6.7).

## SCHRÖDINGER'S WAVE EQUATION

A wave equation capable of describing the motion of an electron was developed by Schrödinger (1925). He introduced a mathematical function $\Psi$ which is a function of the space co-ordinates of the electron and the time. We postulate that $\Psi$ has the form of a solution of the classical wave equation (6.6), so that, restricting the problem to one dimension, the equation for $\Psi$ is

$$\frac{\partial^2 \Psi}{\partial t^2} = \frac{\omega^2}{k^2} \frac{\partial^2 \Psi}{\partial x^2}. \tag{6.8}$$

We now assume that $\Psi$ can be written in terms of time and space functions as

$$\Psi(x, t) = \psi(x)e^{-i\omega t}, \tag{6.9}$$

which is a periodic function of $t$, since

$$e^{-i\omega t} = \cos \omega t - i \sin \omega t,$$

and substituting from eqn (6.9) into eqn (6.8) we have

$$\frac{d^2 \psi}{dx^2} + k^2 \psi = 0. \tag{6.10}$$

Using the de Broglie relation $p = h/\lambda$, $k^2$ can be written as

$$k^2 = \left(\frac{2\pi}{\lambda}\right)^2 = \left(\frac{2\pi p}{h}\right)^2 = \frac{4\pi^2 m^2 v^2}{h^2}, \tag{6.11}$$

where $v$ is the particle velocity. We consider the situation when $v \ll c$ so that $m = m_0$ and the total energy of the electron is given by

$$W = \text{kinetic energy} + \text{potential energy}$$

$$= \tfrac{1}{2}mv^2 + V. \tag{6.12}$$

Hence

$$m^2 v^2 = 2m(W - V)$$

$$k^2 = \frac{8\pi^2 m}{h^2}(W - V), \tag{6.13}$$

and substituting into eqn (6.10) we have

$$\frac{d^2\psi}{dx^2} + \frac{8\pi^2 m}{h^2}(W - V)\psi = 0. \tag{6.14}$$

Equation (6.14) is the *time-independent Schrödinger equation in one dimension*, and has the following features:

(*i*)   It is a time-independent equation and therefore is applicable when the potential energy $V$ is a function of the space co-ordinates but not of the time. (The time-dependent equation is introduced in Problem 6.6.)

(*ii*)   It is, by construction, applicable only in the non-relativistic limit.

(*iii*)   It contains implicitly the de Broglie relation.

(*iv*)   It is a linear equation, i.e. it contains terms involving $\psi$ and its derivatives but does not contain terms independent of $\psi$ or terms involving squares or higher powers of $\psi$ and its derivatives.

The significance of point (*iv*) is that if we have two solutions of the Schrödinger equation, $\psi_1$ and $\psi_2$, then the linearity of the equation ensures that a linear combination of these solutions

$$\psi = a_1\psi_1 + a_2\psi_2$$

is also a solution of the equation for arbitrary values of the constants $a_1$ and $a_2$. We may then satisfy the law of superposition and add wavefunctions just as we add amplitudes when using wave theory to predict interference effects.

## WAVE PACKETS

As a consequence of the law of superposition we can superpose wave-functions in order to construct a wave packet or group, with group velocity $v = d\omega/dk$. This procedure causes localization of the amplitude in a restricted region, and suggests that we associate the localized particle with the wave packet constructed from the superposed wavefunctions. This would imply that the group velocity is equal to the particle velocity, i.e.

$$\frac{d\omega}{dk} = v = \frac{hk}{2\pi m},$$

where we have used eqn (6.11). Re-arranging this expression and integrating we have

$$d\omega = \frac{h}{2\pi m} k \, dk$$

$$\omega = \frac{h^2 k^2}{2\pi m} + \text{constant}.$$

Hence

$$\frac{h\omega}{2\pi} = h\nu = \frac{h^2 k^2}{8\pi^2 m} + \text{constant},$$

$$= \text{kinetic energy} + \text{constant}.$$

We may choose the integration constant in such a way that

$$\frac{h\omega}{2\pi} = h\nu = W, \tag{6.15}$$

where $W$ is the total energy of the electron, so that the relationship between the frequency and total energy for an electron is consistent with Planck's relation $E = h\nu$ for a photon.*

For consistency, we should require that eqn (6.15) also applies at relativistic energies. In this case, it is convenient to use eqns (6.2) and (6.15) to write

$$d\omega = \frac{2\pi}{h} dW, \quad dk = \frac{2\pi}{h} dp,$$

so that the expression for the group velocity can be written as

* The remainder of this section can be omitted on a first reading.

$$U = \frac{dW}{dp}.$$

We can now check that the velocity $U$ of the wave packet is still equal to the particle velocity $v$. For this calculation we can choose the zero of energy, i.e. we can write

$$W = \text{kinetic energy} + \text{potential energy} + \text{rest mass energy},$$

or we can write

$$W = \text{kinetic energy} + \text{potential energy},$$

as in the previous section. This is permissible because the group velocity depends on the derivative of $W$; since the rest mass energy is a constant it will not contribute to $dW$, and we shall get the same answer in both cases. We choose to include the rest mass energy so that $W = E + V$ where $E$ is the total energy, i.e.

$$W = \{c^2p^2 + m_0^2c^4\}^{1/2} + V$$

or

$$c^2p^2 + m_0^2c^4 = W^2 - 2VW + V^2,$$

where we have used eqn (2.11). On differentiating, we have

$$c^2 2p\, dp = 2W\, dW - 2V\, dW$$

so that

$$\frac{dW}{dp} = \frac{c^2p}{W - V} = \frac{c^2p}{\{c^2p^2 + m_0^2c^4\}^{1/2}} = v,$$

where we have used eqns (2.7) and (2.8a) to replace $m_0^2c^4$ by $p^2c^4/v^2 - c^2p^2$.

We can also use eqns (6.11) and (6.15) to determine the wave velocity $v = w/k$ associated with the wavefunction $\Psi$. This is given by

$$u = \frac{2\pi W}{h}\frac{h}{2\pi p} = \frac{W}{p}.$$

In the non-relativistic limit we have

$$W = \tfrac{1}{2}mv^2 + V$$

so that

$$u = \frac{v}{2} + \frac{V}{p}. \tag{6.16}$$

Thus, $u$ depends on the potential energy. For the relativistic calculation we must choose the same zero of energy since $u$ depends directly on $W$. Hence we write

$$W = \{c^2p^2 + m_0^2c^4\}^{1/2} + V - m_0c^2$$

$$u = \frac{W}{p} = \frac{c^2}{v} + \frac{V}{p} - \frac{c^2}{v}\left(1 - \frac{v^2}{c^2}\right)^{1/2}.$$

In the non-relativistic limit $(1 - v^2/c^2)^{1/2} \approx 1 - v^2/2c^2$ and we get the same results as before, while in the high-energy limit $(1 - v^2/c^2)^{1/2} \to 0$ so that

$$u \to \frac{c^2}{v} + \frac{V}{p}. \tag{6.17}$$

Even in this high-energy limit, it must always be true that $v < c$ for a particle of non-zero rest mass, so that for $V \geqslant 0$ we have $u > c$. Thus, for particles of non-zero rest mass we find that $u$ depends on $V$ and may be greater than $c$; in these circumstances it is not possible to assign any real physical significance to the wave velocity $u$. For electromagnetic radiation, the wave velocity in free space is equal to $c$ and hence, from eqn (6.17), the particle velocity for a photon is also equal to $c$, in agreement with Einstein's assumption used in Chapter 2.

## INTERPRETATION OF THE WAVEFUNCTION

Since the wavefunction $\Psi(x, t)$ is a complex quantity whose associated wave velocity $u$ is greater than the velocity of light, neither $\Psi$ nor $u$ can be directly observable quantities. This means that we cannot attribute to $\Psi$ any physical reality but must regard it as a mathematical function from which we may derive physically significant information. One simple real quantity which can be derived from $\Psi$ is the square modulus $|\Psi(x, t)|^2$. An interpretation of this function was first given by Born (1926), who suggested that the quantity $|\Psi(x, t)|^2\, dx$ represents the probability that the electron will be found between $x$ and $x + dx$ if a measurement is made at time $t$ to locate it. Thus $|\Psi(x, t)|^2$ is the probability per unit length of finding the particle in the vicinity of $x$. In the more general three-dimensional case $|\Psi(x, y, z, t)|^2$ is the probability per unit volume or *probability density*. The probability of finding the electron somewhere between $x_1$ and $x_2$ is given by

$$P = \int_{x_1}^{x_2} |\Psi(x, t)|^2\, dx,$$

where the value obtained for $P$ must lie between zero and unity. The maximum value of unity is obtained when we extend the limits of the integral over all space, i.e.

$$1 = \int_{-\infty}^{\infty} |\Psi(x, t)|^2 \, dx, \qquad (6.18)$$

which simply indicates that the probability of finding the electron somewhere in space is unity.

This interpretation in terms of probability is not as alien to classical physics as it might at first appear. There are many circumstances in physics in which we do not attempt to give a strictly causal description of individual particles in an assembly, even though this may in principle be possible, but instead treat the individual behaviour as random and give a statistical description of the behaviour of a large assembly of particles using the laws of probability. Important examples of the procedure occur in the kinetic theory of gases and the law of radioactive decay.

The overwhelming majority of physicists have come to accept this statistical interpretation and have learnt to use the quantum theory to study atomic structure and to make predictions of properties of real physical interest. A few have continued to seek a more concrete picture, or a substructure which can be described by a deterministic theory, as noted in the footnote on page 120.

## SOLUTIONS OF SCHRÖDINGER'S EQUATION FOR SOME SIMPLE CASES

(a) *An electron beam.* We first consider a beam of electrons of infinite width moving along the $x$-direction in a region where there is no potential. A solution of eqn (6.14) with $V = 0$ is

$$\psi(x) = A e^{ikx}, \quad k^2 = 8\pi^2 m W / h^2 \qquad (6.19)$$

so that

$$\Psi(x, t) = A e^{i(kx - \omega t)}$$

$$= A \cos(kx - \omega t) + iA \sin(kx - \omega t).$$

This represents a travelling wave oscillating with angular frequency $\omega = 2\pi W / h$, and moving to the right (i.e. in the direction of increasing $x$) with wave velocity $\omega / k = 2\pi W / k$. It has probability density $P$ given by

$$P = |\Psi(x, t)|^2 = A^* e^{-i(kx - \omega t)} A e^{i(kx - \omega t)} = |A|^2,$$

and hence the number of particles crossing unit area perpendicular to the $x$-axis in unit time is $v|A|^2$. The energy $W = h^2k^2/8\pi^2m$ can take all positive values.

We might expect to determine the coefficient $A$ by requiring that the wavefunction satisfies eqn (6.18), which is known as the *normalization condition*. However, in this case we get

$$\int_{-\infty}^{\infty} |\Psi(x, t)|^2 \, \mathrm{d}x = |A|^2 \int_{-\infty}^{\infty} \mathrm{d}x,$$

and this integral is infinite. This situation arises because the wavefunction represents the idealized situation of a particle moving in an infinitely long beam so that the $x$-coordinate is completely unknown. A realistic wavefunction should describe a group or wave packet of length $\Delta x$ and group velocity $v$, and as long as $\Delta x$ is finite it is possible to determine the normalization integral.

(b) *Electrons in a box.* We now consider electrons constrained to move within a box of side $L$, but still in a region of zero potential. A suitable general solution of eqn (6.14) is

$$\psi(x) = A \sin kx + B \cos kx,$$

but we have to apply the boundary conditions that the wavefunction is zero for $x \leqslant 0$ and $x \geqslant L$, i.e. there are no electrons outside the box. This gives

$$\psi(0) = 0, \quad \therefore B = 0$$

$$\psi(L) = 0, \quad \therefore kL = \pi n,$$

where $n$ is an integer. From the second boundary condition it follows that the energy is now quantized and takes the values

$$W_n = \frac{h^2n^2}{8mL^2}, \quad n = 1, 2, 3, \ldots, \tag{6.20}$$

where we have used eqn (6.13). If $L$ is small the energy levels are widely spaced, but if $L$ is large the energy levels are close together and may almost appear like a continuous spectrum with energy $h^2k^2/8\pi^2m$. The normalization integral can be evaluated in this case and the normalization condition (6.18) gives $A = (2/L)^{\frac{1}{2}}$. The wavefunction

$$\psi_n = \left(\frac{2}{L}\right)^{\frac{1}{2}} \sin\left(\frac{n\pi x}{L}\right),$$

represents standing waves. It should be noted that lowest value of the quantum number $n$ is $n = 1$, since $\psi = 0$ for all $x$ is not a physically interesting solution, and hence the lowest allowed value of the energy is $h^2/8mL^2$ instead of zero. This lowest value is called the *zero-point energy* and its value can be interpreted in terms of the uncertainty principle. The momentum of the particle in this lowest energy state is $p = hk/2\pi = h/2L$. Since the wavefunction represents a standing wave, the associated electron may be moving in either direction so that the uncertainty in momentum is

$$\Delta p_x = 2p = h/L.$$

The electron is confined within the box so that $\Delta x \approx L$ and hence

$$\Delta p_x \, \Delta x \approx h.$$

(c) *Free electrons in metals.* The simplest picture of the motion of conduction electrons in a metal is that they move freely in a region of constant potential. If we treat this region as a three-dimensional box of very large dimension $L$, we simply have to solve the one-dimensional problem three times, and find that the set of energy levels and wavefunctions are given by

$$W_n = \frac{h^2}{8mL^2}\,(n_x^2 + n_y^2 + n_z^2),$$

$$\psi_n = \left(\frac{2}{L}\right)^{3/2} \sin\left(\frac{n_x \pi x}{L}\right) \sin\left(\frac{n_y \pi y}{L}\right) \sin\left(\frac{n_z \pi z}{L}\right).$$

We can now fill up the energy levels corresponding to each set of quantum numbers $n_x n_y n_z$ by putting two electrons of opposite spins in each level, in

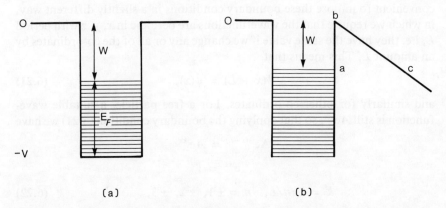

*Figure 6.2.* (a) Energy levels in a metal according to the free electron model. $E_F$ is the Fermi energy and $W$ is the work function. (b) The addition of an electric field giving a potential $-eEx$.

accordance with the exclusion principle. The energy of the highest filled state is called the *Fermi energy* $E_F$. These electrons are all bound in the metal so that we must subtract the constant potential energy $V$, as shown in Fig. 6.2a. The energy required to remove an electron from the highest filled level is now equal to the work function $W$ defined in Chapter 2.

Figure 6.2b shows the effect of applying a strong electric field gradient $E$ to the metal. The potential due to this field is $-eEx$. If the potential is strong enough, electrons can be pulled out of the metal through the potential barrier abc. This phenomenon could not occur classically but can be explained by quantum theory. If $\psi_a$, $\psi_c$ are the wavefunctions at each side of the barrier, the probability $P$ of escape is given approximately by

$$P = |\psi_c/\psi_a|^2,$$

which can be evaluated to give

$$P \approx e^{-\gamma}$$

$$\gamma = \frac{4}{3}\left(\frac{2m}{\hbar^2}\right)^{1/2}\frac{W^{3/2}}{eE}.$$

Increasing the field $E$ decreases $\gamma$ and hence $P$ increases. Conversely, increasing the work function $W$ increases $\gamma$ and so decreases $P$. The escape of electrons by this means is known as *quantum mechanical tunnelling* (this process is used again in Chapter 7 to explain the phenomenon of $\alpha$-decay).

(d) *Electrons in a periodic potential.* In the previous example, the use of boundary conditions on $\psi$ at the edge of the box causes quantization of the energy and makes it possible to normalize the wavefunctions. It is often convenient to impose these boundary conditions in a slightly different way, in which we require that the wavefunctions are periodic in $x$, $y$, $z$ with period $L$, i.e. they have the same value if we change any or all of the co-ordinates by an amount $L$. This means that

$$\psi(x + L) = \psi(x), \tag{6.21}$$

and similarly for other co-ordinates. For a free particle, a suitable wavefunction is still $Ae^{ikx}$, so that applying the boundary condition (6.21) we have

$$Ae^{ik(x+L)} = Ae^{ikx}$$

$$\therefore e^{ikL} = 1$$

$$k = 2\pi n/L, \quad n = \pm 1, \pm 2, \pm 3, \ldots. \tag{6.22}$$

We can determine $A$ by requiring that the normalization condition (6.18) is satisfied over the range $x = 0$ to $x = L$. The wavefunction represents a travelling wave again and is given by

$$\psi_n = \frac{1}{L^{3/2}} \exp \left\{ i \frac{2\pi}{L} (n_x x + n_y y + n_z z) \right\},$$

and the energy by

$$W_n = \frac{h^2}{2mL^{3/2}} (n_x^2 + n_y^2 + n_z^2). \tag{6.23}$$

This method may be used to improve the description of the motion of electrons in crystalline solids. The presence of the ions in the crystal structure means that the free electrons do not move in a constant potential but instead move in a periodic potential which repeats itself with a period $a$, where $a$ is

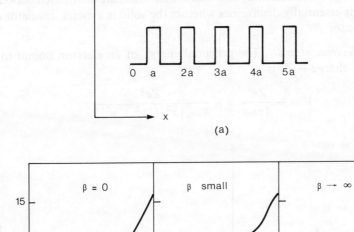

(a)

(b)

*Figure 6.3.* (a) A version of the periodic potential experienced by electrons in a metal due to the presence of the ions. (b) The variation of the electron energies $W$ as a function of $k$ for the three cases, $\beta = 0$, $\beta = $ finite and small, and $\beta \to \infty$, as discussed in the text.

the spacing of the ions. Such a potential can be represented by the function shown in Fig. 6.3a. We use the symbol $\beta$ to represent the height and width of the potential spike around each value of $a$, in such a way that if $\beta = 0$ there is no spike at all and $V$ is constant, while if $\beta \to \infty$ the potential spike is very narrow and infinitely high. Now, if $\beta = 0$ the electron is free to move through the lattice so that we get a travelling wave with energy $W \propto k^2$. If $\beta \to \infty$ the electrons are effectively held within a box of side $a$ so that we get standing waves with quantized energies $W_n = h^2n^2/8ma^2$. In an intermediate situation when $\beta$ is small but non-zero, it can be shown that there is a transition between travelling waves for $k \neq n\pi/a$ and standing waves for $k = \pm n\pi/a$. The three situations are illustrated in Fig. 6.3b, where $W$ is plotted as a function of $k$. The important result obtained with this simple model is that there can be an *energy gap* at $k = \pm n\pi/a$ between *bands* of allowed energy levels. The way in which the electrons are distributed between these bands essentially determines whether the solid is a metal, insulator or semiconductor.

*(e) One-electron atoms.* The potential energy of an electron bound to a nucleus of charge $+Ze$ is

$$V = -\frac{Ze^2}{4\pi\epsilon_0 r} = -\frac{Ze^2}{4\pi\epsilon_0\{x^2 + y^2 + z^2\}^{1/2}}.$$

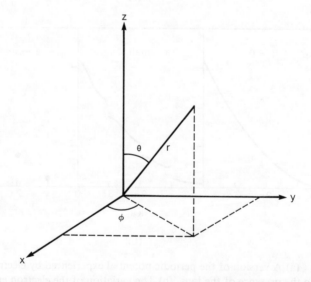

*Figure 6.4.* Definition of the spherical polar co-ordinates $r$, $\theta$, $\phi$ in terms of the cartesian co-ordinates $x$, $y$, $z$.

It is clear that we must treat this problem in three dimensions, and also that the form of the potential is much simpler when expressed in terms of $r$ than in terms of $x$, $y$, $z$. For these reasons we discuss this example in terms of the spherical polar co-ordinates $(r, \theta, \phi)$ defined in Fig. 6.4. The boundary conditions on $\psi(r, \theta, \phi)$ are that it:

(i)   is finite at $r = 0$;

(ii)   tends to zero as $r \to \infty$;

(iii)   is finite for all $\theta$ including $\theta = 0, \pi$;

(iv)   has the same value when $\phi$ is changed to $\phi + 2\pi$.

These boundary conditions ensure that the mathematical expression we get by solving the equation represents a physically realistic solution. For example, conditions (i) and (iii) imply that the probability of finding the electron is finite everywhere, while condition (ii) implies that the probability of finding the electron at a large distance from the atom is negligibly small, i.e. the electron is indeed bound to the atom. Condition (iv) is necessary because $\phi$ and $\phi + 2\pi$ represent the same point in space, as can be seen from Fig. 6.4. The first two boundary conditions lead to quantization of energy and introduce the principal quantum number $n$, the third boundary condition leads to quantization of angular momentum and introduces the orbital quantum number $l$, and the fourth condition introduces the quantum number $m_l$. The formula for the energy is exactly that given by the Bohr theory and the values taken by the quantum numbers are those deduced from experiment.

The first few solutions of the Schrödinger equation for one-electron atoms are given in Table 6.1. For electrons in states with $l \neq 0$, this table shows that the corresponding wavefunctions depend on $\theta$ and $\phi$ as well as $r$, so that the normalization condition must be written as an integral over the whole volume using the volume element $\mathrm{d}v$, which in spherical polar co-ordinates is $r^2 \sin\theta \, \mathrm{d}r \, \mathrm{d}\theta \, \mathrm{d}\phi$. This gives

$$1 = \int_{\mathrm{vol}} |\psi|^2 \, \mathrm{d}v \tag{6.23a}$$

$$= \int_0^\infty \int_0^\pi \int_0^{2\pi} |\psi(r, \theta, \phi)|^2 \, r^2 \, \mathrm{d}r \, \sin\theta \, \mathrm{d}\theta \, \mathrm{d}\phi. \tag{6.24b}$$

The interpretation of these wavefunctions is again obtained through probability arguments. It can be seen from Table 6.1 that the wavefunction $\psi(r, \theta, \phi)$ can be written as products of a function of $r$, a function of $\theta$ and a

*Table 6.1.* Wavefunction for the electron in a one-electron atom

| $n$ | $l$ | $m_l$ | $\psi(r, \theta, \phi)$ |
|---|---|---|---|
| 1 | 0 | 0 | $\psi = \dfrac{1}{\sqrt{\pi}}\left(\dfrac{Z}{a_0}\right)^{3/2} e^{-Zr/a_0}$ |
| 2 | 0 | 0 | $\psi = \dfrac{1}{4\sqrt{2\pi}}\left(\dfrac{Z}{a_0}\right)^{3/2}\left(2 - \dfrac{Zr}{a_0}\right)e^{-Zr/a_0}$ |
| 2 | 1 | 0 | $\psi = \dfrac{1}{4\sqrt{2\pi}}\left(\dfrac{Z}{a_0}\right)^{3/2}\left(\dfrac{Zr}{a_0}\right)e^{-Zr/2a_0}\cos\theta$ |
| 2 | 1 | $\pm 1$ | $\psi = \dfrac{1}{8\sqrt{\pi}}\left(\dfrac{Z}{a_0}\right)^{3/2}\left(\dfrac{Zr}{a_0}\right)e^{-Zr/2a_0}\sin\theta\, e^{\pm i\phi}$ |
| 3 | 0 | 0 | $\psi = \dfrac{1}{81\sqrt{3\pi}}\left(\dfrac{Z}{a_0}\right)^{3/2}\left(27 - 18\dfrac{Zr}{a_0} + 2\dfrac{Z^2 r^2}{a_0}\right)e^{-Zr/3a_0}$ |

The constant $a_0$ has dimensions of length and is equal to $h^2\epsilon_0/\pi\mu Ze^2$, i.e. it is the Bohr radius for the orbit $n = 1$.

function of $\phi$, and, further, that the function of $\phi$ is of the form $e^{\pm im\phi}$. This means that the probability density is independent of $\phi$ and can be written as

$$P = |\psi_{nl\,m_l}(r, \theta, \phi)|^2 = |R_{nl}(r)|^2 |\theta_{l\,m_l}(\theta)|^2. \qquad (6.25)$$

We first consider the states with $l = 0$ which are independent of $\theta$, so that $P$ is spherically symmetric. We can interpret the quantity $4\pi r^2|\psi|^2\,dr = 4\pi r^2\,dr\,P$ as the probability of finding the electron in a shell of radius $r$ and volume $4\pi r^2\,dr$ around the nucleus. The quantity $4\pi r^2|\psi|^2$ is plotted in Fig. 6.5 for some spherically symmetric states of hydrogen. In the case of the ground state, $n = 1, l = 0$, we have marked the position of the Bohr radius, and it can be seen that, although the maximum probability occurs at the Bohr radius, there is a substantial probability for finding the electron elsewhere.

Some of the functions $|\theta|^2$ are shown in Fig. 6.6. It is clear that there can be remarkable variation in the distribution of the probability and hence in the distribution of charge. This has an important effect on the formation of chemical bonds between atoms.

It is evident that there is no longer a uniquely defined radius for an electron in a given state. It is often useful to calculate the *mean square radius* $\langle r^2 \rangle$

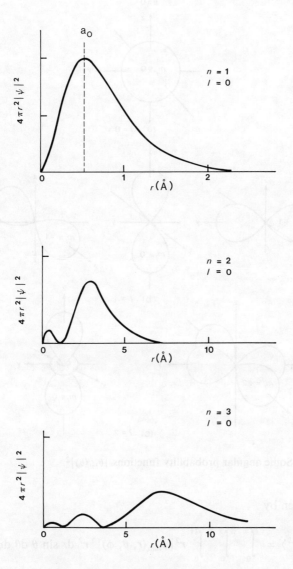

*Figure 6.5.* Some radial probability functions $4\pi r^2 \, |R_{nl}(r)|^2$ for the hydrogen atom.

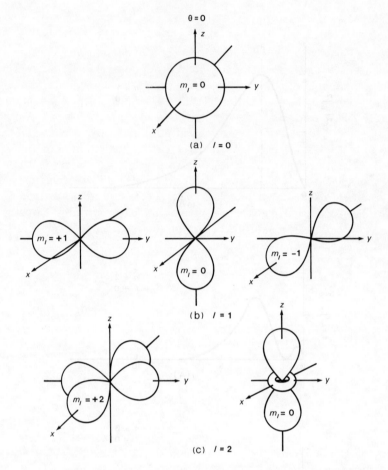

*Figure 6.6.*   Some angular probability functions $|\theta_{lm_l}(\theta)|^2$.

which is given by

$$\langle r^2 \rangle = \int_0^\infty \int_0^\pi \int_0^{2\pi} r^2 |\psi_{nlm_l}(r, \theta, \phi)|^2 \, r^2 \, dr \, \sin\theta \, d\theta \, d\phi \quad (6.26a)$$

$$= \int_0^\infty r^2 |R_{nl}(r)|^2 r^2 \, dr. \quad (6.26b)$$

The root mean square (rms) radius $\langle r^2 \rangle^{1/2}$ clearly has dimensions of length and can be used as a measure of the size of the atom.

## FURTHER DEVELOPMENT OF THE QUANTUM THEORY

With the aid of Schrödinger's time-independent equation we have been able to determine the allowed energies and corresponding wavefunctions in a number of simple situations, and we have seen that the application of the appropriate boundary conditions leads to the introduction of quantum numbers. However, we often want to know much more about the system than just the energy — we want to know the probability for a transition from one state to another due to different types of excitation. For this it is necessary to associate each physical quantity with a quantum mechanical operator in such a way that the result of operating on a wavefunction yields the required information. The details of this procedure are beyond the scope of this book but we can give one simple example to show how it works. In spherical polar co-ordinates the operator for the $z$-component of orbital angular momentum is found to be

$$- i \frac{h}{2\pi} \frac{\partial}{\partial \phi}.$$

Now if we operate with this operator on the wavefunctions for the one-electron atom given in Table 6.1 we get zero if $m_l = 0$, or $\pm 1$ if $m_l \pm 1$. The result can be written in general form as

$$- i \frac{h}{2\pi} \frac{\partial}{\partial \phi} \psi(r, \theta, \phi) = - \frac{ih}{2\pi} (im_l)\psi(r, \theta, \phi)$$

$$= m_l \frac{h}{2\pi} \psi(r, \theta, \phi).$$

Thus, the magnitude of the $z$-component of orbital angular momentum is $m_l(h/2\pi)$ in accordance with the definition of $m_l$ given in Chapter 5.

## FORMATION OF CHEMICAL BONDS AND MOLECULES

In the formation of chemical bonds, the electrons of closed inner shells are not much involved. Chemical bonding takes place between the outer or *valence* electrons which are more loosely bound. For example, sodium has full $n = 1$ and $n = 2$ shells and a single valence electron in a 3s state. Chlorine has one vacancy in the 3p state and so behaves with valency one rather than 7.

A case of particular interest in organic chemistry and in the formation of biological molecules is the behaviour of the carbon atom. Carbon has six electrons, two in the closed 1s shell, and the remainder in the 2s and 2p states.

We denote the three different spatial forms of the 2p wavefunction as $2p_1$, $2p_2$, $2p_3$. It is possible to make linear combinations of the 2s and 2p wavefunctions. This is known as *hybridization*.

In *tetragonal* hybridization we make four linear combinations whose centres of charge are displaced towards the four vertices of a tetrahedron. These new wavefunctions are

$$\psi_1 = \tfrac{1}{2}(\phi_s + \phi_{p_1} + \phi_{p_2} + \phi_{p_3})$$

$$\psi_2 = \tfrac{1}{2}(\phi_s + \phi_{p_1} - \phi_{p_2} - \phi_{p_3})$$

$$\psi_3 = \tfrac{1}{2}(\phi_s - \phi_{p_1} + \phi_{p_2} - \phi_{p_3})$$

$$\psi_4 = \tfrac{1}{2}(\phi_s - \phi_{p_1} - \phi_{p_2} + \phi_{p_3}).$$

Each of the four wavefunctions can take part in forming a bond. If each bond is with hydrogen we get the molecule $CH_4$, where the carbon atom sits in the centre of a tetrahedron whose vertices are occupied by the hydrogen atoms. The density distribution of the electrons in the carbon atoms is illustrated in Fig. 6.7a.

(a)

(b)

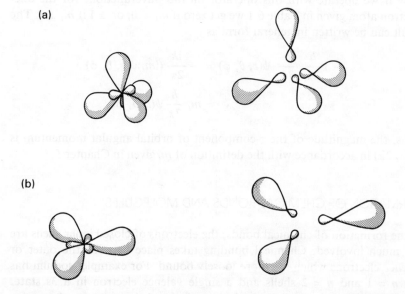

*Figure 6.7.* (a) The density distribution of the four electrons giving the $sp^3$ hybridization of carbon, with an expanded view on the right. (b) The density distribution of the three electrons giving the $sp^2$ hybridization of carbon, with an expanded view on the right. (From H. Haken and H. C. Wolf, *Atomic and Quantum Physics*. Springer-Verlag, 1984.)

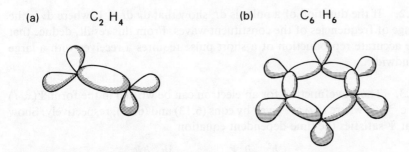

(a) $C_2H_4$  (b) $C_6H_6$

*Figure 6.8.* (a) A possible density distribution of the hybridized electrons in ethene, $C_2H_4$. (b) Electron density distribution in benzene, $C_6H_6$. (After H. Haken and H.C. Wolf, *Atomic and Quantum Physics*. Springer-Verlag, 1984.)

Another possibility is *trigonal* hybridization in which linear combinations can be made of the form

$$\psi_1 = \frac{1}{3}(\phi_s + 2\phi_{p_1})$$

$$\psi_2 = \frac{1}{3}\left(\phi_s + \sqrt{\frac{3}{2}}\,\phi_{p_2} - \sqrt{\frac{1}{2}}\,\phi_{p_1}\right)$$

$$\psi_3 = \frac{1}{3}\left(\phi_s - \sqrt{\frac{3}{2}}\,\phi_{p_2} - \sqrt{\frac{1}{2}}\,\phi_{p_1}\right)$$

Hydrogen–carbon bonds are formed by $\psi_2$ and $\psi_3$, while $\psi_1$ forms a carbon–carbon bond. The density distribution of the electrons in the carbon atoms is illustrated in Fig. 6.7b.

The unused $p_3$ function can be used to form another bond between two carbon atoms, as shown in Fig. 6.8 for ethene ($C_2H_4$) and benzene ($C_6H_6$).

## PROBLEMS

**6.1.** Using the definition of angular frequency $\omega$ show that

$$\frac{d\omega}{d\lambda} = \frac{-2\pi u}{\lambda^2} + \frac{2\pi}{\lambda}\frac{du}{d\lambda}.$$

Hence show that the group velocity is given by

$$U = u - \lambda\frac{du}{d\lambda}.$$

**\*6.2.** If the duration of a pulse is $dt$, show that $d\nu \, dt \approx 1$, where $d\nu$ is the range of frequencies of the constituent waves. From this result, deduce that the accurate reproduction of a short pulse requires a receiver with a large bandwidth.

**\*6.3.** The wavefunction for an electron can be written in the form $\Psi(x, t)$ $= e^{i(kx - \omega t)}$, where $k$, $\omega$ are given by eqns (6.13) and (6.15), respectively. Show that $\Psi$ satisfies the time-dependent equation

$$-\frac{h^2}{8\pi^2 m}\frac{\partial^2 \Psi}{\partial x^2} + V\psi = \frac{ih}{2\pi}\frac{\partial \Psi}{\partial t}.$$

**6.4.** Using the functions $\psi(r, \theta, \phi)$ given in Table 6.1, examine the behaviour of $|\psi|^2$ and $4\pi r^2 |\psi|^2$. Consider the possible ways of representing this behaviour by means of graphs and three-dimensional models.

**\*6.5.** Using the wavefunctions given in Table 6.1 compare the root mean square radii $\langle r^2 \rangle^{1/2}$ for the states: $n = 1, l = 0$; $n = 2, l = 0$; $n = 3, l = 0$.

# 7 | Physics of the Nucleus

In Chapter 4 we examined some preliminary information about the atomic nucleus obtained from experiments on $\alpha$-particle scattering. This indicated that the nucleus contains almost all the mass of the atom in a volume so small that the density of the nucleus is of the order of $10^{12}$ times higher than the density of matter in bulk. We have also noted from many different phenomena that the atomic number $Z$, which determines the charge on the nucleus, is a quantity of fundamental significance. A more subtle point is that in the study of atomic properties we have been able to regard the nucleus as essentially inert, which implies that energy changes of the order of eV, and possible keV, do not disturb the nucleus. It is therefore important to determine whether the general principles and the quantum theory we have used to interpret atomic properties also apply to nuclear properties.

## CONSTITUENTS OF NUCLEI

Nuclei are composed of protons and neutrons. The proton is the nucleus of the lightest and most common isotope of hydrogen; it has a mass of 1.00783 amu, an intrinsic spin of $\frac{1}{2}(h/2\pi)$ and a charge of $+e$, and obeys the exclusion principle. The neutron also has a spin of $\frac{1}{2}(h/2\pi)$ and obeys the exclusion principle, but it has no charge and has a mass of 1.00867 amu. Protons and neutrons are often referred to collectively as *nucleons*. They are fundamental particles, in the same sense as electrons and the other particles listed in Table 7.1.

Earlier ideas assumed that nuclei were composed of protons and electrons. For a nucleus of atomic number $Z$ and mass number $A$, the nucleus would then be composed of $A$ protons and $A - Z$ electrons. There are two objections to this assumption. One arises from the uncertainty principle (see Problem 7.7). The other is connected with the spin of the nucleus and its constituents. Protons and neutrons have spin $\frac{1}{2}$, so that if $Z + N$ is even the total

Table 7.1. Properties of some fundamental particles

| Particle | | Mass in units of electron mass | Charge in units of electron charge | Spin in units of $(h/2\pi)$ | Antiparticle | Interactions |
|---|---|---|---|---|---|---|
| **Leptons** | | | | | | |
| Neutrino | $\nu_e, \nu_\mu$ | 0 | 0 | $\frac{1}{2}$ | $\bar{\nu}_e, \bar{\nu}_\mu$ (anti-neutrino) | Weak |
| Electron | $e^-$ | 1 | $-1$ | $\frac{1}{2}$ | $e^+$ (positron) | ⎫ Weak, electromagnetic |
| Muon | $\mu^-$ | 206.8 | $-1$ | $\frac{1}{2}$ | $\mu^+$ | ⎭ |
| **Mesons** | | | | | | |
| Pion | $\pi^0$ | 264.2 | 0 | 0 | Itself | |
| | $\pi^+$ | 273.2 | $+1$ | 0 | $\pi^-$ | Weak, electromagnetic |
| Kaon | $K^+$ | 966.6 | $+1$ | 0 | $K^-$ | and strong |
| | $K^0$ | 974 | 0 | 0 | $\bar{K}^0$ | |
| Eta | $\eta$ | 1072 | 0 | 0 | Itself | |
| **Baryons** | | | | | | |
| Proton | $p$ | 1836.1 | $+1$ | $\frac{1}{2}$ | $\bar{p}$ (anti-proton) | |
| Neutron | $n$ | 1838.7 | 0 | $\frac{1}{2}$ | $\bar{n}$ (anti-neutron) | |
| Lambda | $\Lambda$ | 2182.8 | 0 | $\frac{1}{2}$ | $\bar{\Lambda}$ | |
| Sigma | $\Sigma^+$ | 2327.7 | $+1$ | $\frac{1}{2}$ | $\bar{\Sigma}^-$ | Weak, electromagnetic |
| | $\Sigma^0$ | 2331.8 | 0 | $\frac{1}{2}$ | $\bar{\Sigma}^0$ | and strong |
| | $\Sigma^-$ | 2340.5 | $-1$ | $\frac{1}{2}$ | $\bar{\Sigma}^+$ | |
| Xi | $\Xi^0$ | 2565 | 0 | $\frac{1}{2}$ | $\bar{\Xi}^0$ | |
| | $\Xi^-$ | 2580 | $-1$ | $\frac{1}{2}$ | $\bar{\Xi}^+$ | |
| Omega | $\Omega^-$ | 3300 | $-1$ | $\frac{3}{2}$ | $\bar{\Omega}^+$ | |

spin is an integer, even if both $Z$ and $N$ are odd. Electrons also have spin $\frac{1}{2}$ so that the proton plus electron model would give $2A - Z$ particles of spin $\frac{1}{2}$ so that if $A$ is even and $Z$ is odd then $2A - Z$ is also odd. This result is inconsistent with the values given in Table 7.2.

Since the masses of the proton and neutron are so close to one atomic mass unit and the electron mass is so small by comparison, we can now understand why isotopic masses are in all cases very close to an integer value in amu. It is usual to quote the mass of a given isotope in amu. The relation between the atomic mass $M_a(Z, A)$ of an isotope with mass number $A$ and atomic number $Z$ and the nuclear mass $M_{nuc}(Z, A)$ is given by

$$M_a(Z, A) = M_{nuc}(Z, A) + Zm_e,$$

where $m_e$ is the electron mass and the binding of electrons in atoms has been neglected. Further, we see that the value of the atomic number $Z$, and hence the position in the periodic table, is determined by the number of protons in the nucleus, while the total number of nucleons determines the mass number $A$. For a given $Z$, a change in the neutron number $N = A - Z$ causes a change in the nuclear and isotopic mass of approximately integer values of amu. For light elements, the stable isotopes have $Z \approx N$, while for heavier nuclei the stable isotopes have $Z < N$. In heavy nuclei, the purely electrostatic Coulomb repulsion between the protons makes the isotopes with $Z = N$ unstable.

If we plot the nuclei and reasonably long-lived radioactive nuclei against $Z$ and $N$ they fall in a narrow band which follows the line $Z = N$ only up to $Z \leqslant 20$, as shown in Fig. 7.1. Beyond $Z = 20$ the band curves away from the $Z = N$ line towards the $N$ axis. Nuclei falling in this band are said to be in "the valley of stability". Nuclei falling well outside the band may be formed in nuclear reactions, but have very short lifetimes.

## NUCLEAR SIZES AND SHAPES

In Chapter 4 we assumed that the nucleus could be regarded as a sphere of constant density, i.e. the density of the nucleus is assumed to have a constant value, say $\rho_0$, up to some radius $R$ where it falls sharply to zero, as shown in Fig. 7.2a. In practice, it is found that this assumption is too crude to give an adequate description of the data, and a more realistic form, such as that shown in Fig. 7.2b must be used. The functions shown in Fig. 7.2 represent the distribution of probability density; they are called *density distributions* and are denoted by the symbol $\rho(r)$.

In Fig. 7.2b the nucleus is no longer assumed to have a sharp surface but has a diffuse surface with a *definite surface thickness t*, which is defined as

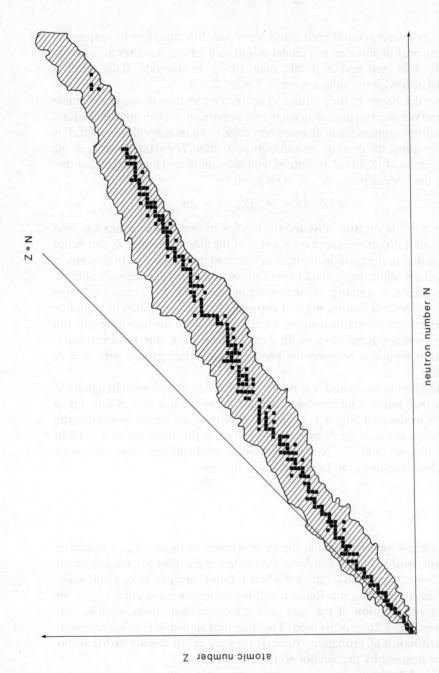

Figure 7.1. Stable and long-lived radioactive nuclei plotted against $Z$ and $N$ showing the effect of a neutron excess beyond the very light nuclei and the trend of the "valley of stability". Black squares indicate stable nuclei while the shaded region indicates where long-lived radioactive nuclei occur.

atomic number Z

neutron number N

Z = N

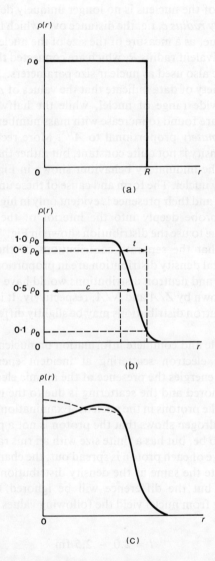

*Figure 7.2.* Possible forms for the density distribution $\rho(r)$ of matter in the nucleus as functions of the distance $r$ from the centre of the nucleus.

the distance over which the density falls from 90% to 10% of its central value. The radius of the nucleus is no longer uniquely defined, although we can use the *halfway radius c*, i.e. the distance over which the density drops to half its central value, as a measure of the size of the nucleus. The rms radius $\langle r^2 \rangle^{1/2}$ and the equivalent radius $R$, which are connected through the relation $R^2 = 5\langle r^2 \rangle/3$, are also used as nuclear size parameters.

Studies of a variety of data indicate that the values of $\rho_0$ and $t$ are roughly constant over a wide range of nuclei, while the halfway radius and the equivalent radius are found to increase with mass number in such a way that both are *approximately* proportional to $A^{1/3}$. More recent studies suggest that the central density is not quite constant, but rather that the density in the inner region has the undulatory behaviour shown in Fig. 7.2c, at least for medium and heavy nuclei. The form and cause of these undulations is not yet clearly established and their presence is evident only in high-energy scattering processes which probe deeply into the interior of the nucleus. For this reason, we continue to use the distribution shown in Fig. 7.2b. Until recently it was assumed that the separate contributions of the protons and the neutrons to the total density distribution are in proportion to their numbers, so that the proton and neutron distributions would have the same shape but would be scaled down by $Z/A$ and $N/A$, respectively. It is now believed that the proton and neutron distributions may be slightly different in the extreme surface region.

The most reliable and complete information on nuclear sizes comes from studies of elastic electron scattering at incident energies above about 50 MeV. At these energies the presence of the atomic electrons around each nucleus can be ignored and the scattering is due to the interaction between the electrons and the protons in the nucleus. Examination of the scattering of electrons from hydrogen shows that the proton is not a point charge, as the electron appears to be, but has a finite size with an rms radius of $\sim 0.8$ fm.* (Because the charge of each proton is spread out, the charge distribution of a nucleus is not quite the same as the density distribution of the protons, as discussed above, but the difference will be ignored here.) Analyses of electron scattering from nuclei yield the following values for the nuclear size parameters:

$$t \approx 2.0 - 2.5 \text{ fm}$$

$$c \approx 1.0 \, A^{1/3} \text{ fm} \qquad\qquad (7.1)$$

$$R \approx 1.2 \, A^{1/2} \text{ fm}.$$

---

* The femtometer (fm) is defined in Chapter 1.

We have so far assumed that nuclei are spherical, but this is not generally true, and many nuclei are ellipsoidal in shape. One quantity which measures the deviation of the nuclear charge distribution from spherical shape is the electric quadrupole moment which is zero for spherical nuclei. It can be estimated from electron scattering and can be measured more accurately in a variety of processes which involve interaction between the nucleus and an applied field or with the field due to the atomic electrons. Figure 7.3 shows

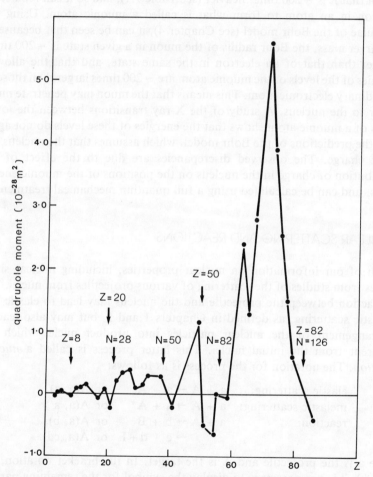

*Figure 7.3* Values of the nuclear quadrupole moment plotted as a function of atomic number $Z$. For each element the value for the most abundant isotope is plotted, if known.

the variation of the electric quadrupole moment with mass number. The quadrupole moment is defined so that it has units of area. The nuclear units of area are the *barn* and the *millibarn*, such that

$$1 \text{ barn} = 10^{-28} \text{ m}^2, \quad 10 \text{ mb} = 1 \text{ fm}^2.$$

Further information about the shape and size of the nuclear charge distribution is obtained from measurements of X-ray transitions of muonic atoms. The negative muon resembles the electron in almost every respect except that it is ~200 times heavier (see Table 7.1), and so it can replace an electron in an atom to form what is called a muonic atom. Using the formulae of the Bohr model (see Chapter 4), it can be seen that because of the larger mass, the Bohr radius of the muon in a given state is ~200 times smaller than that of an electron in the same state, and that the allowed energies of the levels of the muonic atom are ~200 times larger than those of an ordinary electronic atom. This means that the muon may penetrate much closer to the nucleus. A study of the X-ray transitions between the lower levels of a muonic atom shows that the energies of these levels do not agree with the predictions of the Bohr model, which assumes that the nucleus is a point charge. The observed discrepancies are due to the effect of the distribution of charge in the nucleus on the positions of the muonic energy levels, and can be calculated using a full quantum mechanical treatment.

## NUCLEAR SCATTERING AND REACTIONS

Much of our information on nuclear properties, including nuclear sizes, comes from studies of the scattering of various projectiles from nuclei. The interaction between the projectile and the nucleus may lead to elastic and inelastic scattering, as defined in Chapters 1 and 5, but may also lead to rearrangement of the nuclear particles into product nuclei which are different from the initial nuclei. This latter process is called a *nuclear reaction*. The notation for the process is as follows:

$$\text{elastic scattering:} \quad a + A \rightarrow a + A \quad \text{or A}(a, a)$$
$$\text{inelastic scattering:} \quad a + A \rightarrow a + A^* \quad \text{or A}(a, a')$$
$$\text{reaction:} \quad a + A \rightarrow b + B \quad \text{or A}(a, b)$$
$$\rightarrow c + d + E \quad \text{or A}(a, cd).$$

Here a is the projectile and A is the target. In the bracket notation, i.e. A(a, b), it is not necessary to display the symbol for the remaining particle because we can deduce its identity from conservation of charge and conservation of nucleon number, provided that the mass number and atomic number of all the other particles are known.

The data for a given process at a fixed incident energy are usually presented in terms of the *differential cross-section* $d\sigma/d\Omega$ and the *total cross-section* $\sigma$. We define $d\sigma$ as the number of particles scattered per unit time through a scattering angle between $\theta$ and $d\theta$ into a solid angle $d\Omega = 2\pi \sin \theta \, d\theta$* if the incident beam carries one particle per second and there is one target nucleus per unit area. The differential cross-section then gives the number scattered divided by the solid angle $d\Omega$, as a function of the scattering angle $\theta$. The integral of the differential cross-section over all solid angles is the total cross-section, i.e.

$$\sigma = 2\pi \int_0^\pi \frac{d\sigma}{d\Omega} \sin \theta \, d\theta. \qquad (7.2)$$

This quantity is clearly not a function of scattering angle but has one value for a given process at a fixed energy, and represents the probability for that process to occur if the incident beam carries one particle per second and the target contains one nucleus per unit area. The units of the total cross-section are those of area and the units of the differential cross-section are area divided by the unit of solid angle (steradian). Typical differential cross-sections are shown in Fig. 7.4.

The total cross-section for a given process is often plotted as a function of the projectile energy. This function is then called the *excitation function* for the process. The total cross-section for all processes which can occur for a given projectile and energy is the sum of the total cross-section $\sigma_{el}$ for elastic scattering and the total cross-section $\sigma_{abs}$ for all other processes which lead to depletion of the elastically scattered beam, i.e.

$$\sigma = \sigma_{el} + \sigma_{abs}. \qquad (7.3)$$

The value of the total cross-section is independent of the frame of reference since the probability of a process must be independent of our description of the colliding system. The differential cross-section and the scattering angle are dependent on the frame of reference, and the data may be quoted relative to the *laboratory frame*, in which the target nucleus is initially at rest, or to the *centre of mass frame*, in which the centre of mass of the whole system (projectile plus target) is at rest (see Chapter 1).

We have already seen that electron scattering can be used to study the nuclear charge distribution, but the electron is not a nuclear particle and cannot give any information about the nuclear force. In contrast, scattering

---

* For simplicity we have assumed that the scattering is independent of the azimuthal angle $\phi$, i.e. it is axially symmetric. If this is not the case the solid angle is given by $d\Omega = \sin \theta \, d\theta \, d\phi$ and the differential cross-section is a function of $\phi$ as well as $\theta$.

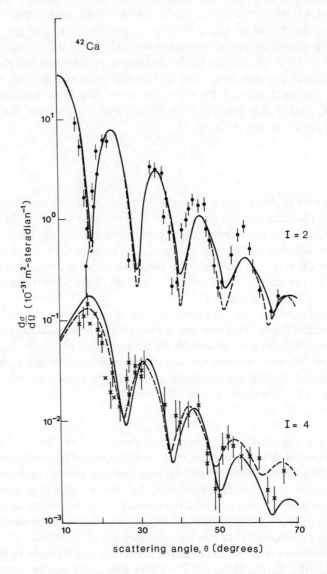

*Figure 7.4.* Examples of differential cross-sections for the scattering of (a) medium energy $\alpha$-particles and

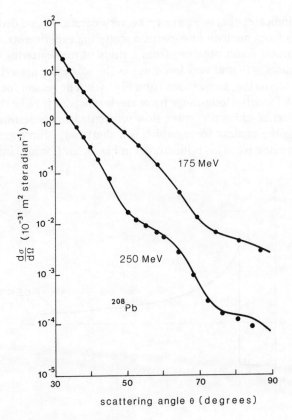

*Figure 7.4.* (b) high-energy electrons.

of nucleons from nuclei gives information about the interaction of a nucleon and the target nucleus, scattering of $\alpha$-particles gives information on the interaction of an $\alpha$-particle and the nucleus, and so on. If these interactions are represented by a potential, it is found that the potential must have a rather similar shape to the charge distribution, and in particular must have a diffuse surface. The magnitude of the central value and the halfway radius are found to depend on the nature of the projectile and its energy. For medium energy projectiles with energies $< 100$ MeV it is roughly true that the halfway radius for $\alpha$-particle scattering is greater than the halfway radius for nucleon scattering, which in turn is greater than the halfway radius of the charge distribution, while at high energies the radii all tend to the same value. This result contains some important information about the nuclear force, but it also indicates that we have to be very careful about deductions of nuclear sizes from nucleon or $\alpha$-particle scattering experiments.

One important result obtained from a study of the scattering of charged nuclear particles is that at very low energies the scattering always reduces to Rutherford scattering, as was seen from Fig. 4.3. The reason for this is that the Coulomb force is a long-range force and is repulsive for like charges, and so forms a barrier which prevents a slow projectile from penetrating into the region where the nuclear force, which is a short-range force and generally attractive, is effective. This is illustrated in Fig. 7.5a. It is also interesting to

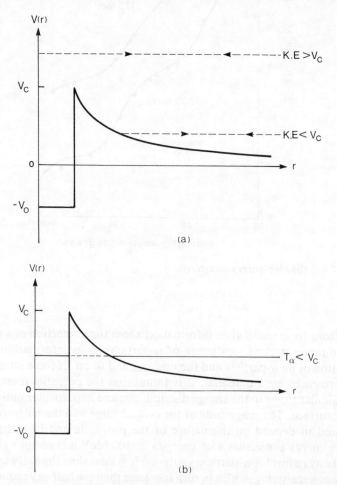

*Figure 7.5.* The effect of the Coulomb barrier on (a) low-energy $\alpha$-particle scattering and (b) $\alpha$-particle decay.

note that, in the special case of Rutherford scattering, a theoretical prediction derived from the Schrödinger equation yields exactly the same formula as that obtained from classical theory. Such agreement between the predictions of classical and quantum theory is rather uncommon, but the agreement in the case of Rutherford scattering, together with the observation that there is no evidence for a breakdown of Coulomb's law in electron scattering, confirms the conclusion stated in Chapter 4 that the departure from Rutherford scattering at increasing energies is due to the existence of the nuclear force.

## NUCLEAR ENERGY LEVELS

Studies of inelastic scattering and reactions show that nuclei have a discrete spectrum of excited states. A typical energy level diagram for a light nucleus is shown in Fig. 7.6, where each horizontal line represents an energy level in the usual way. The zero of energy is taken to be the energy of the ground state, so that the value of the energy associated with each line denotes the excitation energy of that state above the ground state. It can be seen that the spacing of the low-lying levels is of the order of MeV but the higher levels are much more closely spaced. When the nucleus is in an excited state it may give up its energy and return to the ground state through the emission of a photon. The energy of the photon is determined by the difference between the energies of the states, as given by Bohr's second postulate and eqn (4.13), and the magnitude is such that the photon is in the $\gamma$-ray region of the electromagnetic spectrum. Figure 7.6 also shows the minimum energy required to cause the nucleus to break up into various components. Since these values are all above the ground state energy, it is clear that the nucleus in its ground state is stable against these decays. However, those excited states which are above the thresholds can break up in various ways. This means that the nucleus does not remain indefinitely in an excited state; thus the lifetime $\tau$ of the state is finite and may be short or long depending on the probability of decay, the selection rules and the number of possible decay modes. It can be seen from the uncertainty principle in the form $\Delta E \, \Delta t \approx h$ that if the lifetime is finite the energy of the state cannot be completely sharp, but instead has a *width* $\Gamma$ or energy given by eqn (3.32d)* as

$$\Gamma = h/2\pi\tau. \tag{7.4}$$

* This argument applies quite generally to quantized systems. It leads to a *natural line width* for the electromagnetic radiation emitted in transition between levels.

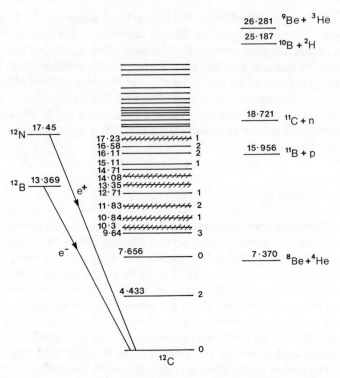

*Figure 7.6.*   The energy level diagram for the nucleus $^{12}$C. The low-lying levels are labelled with the excitation energy and the nuclear spin. The cross-hatched levels have large widths. On the right of the nuclear level diagram are shown the separation energies for break-up into various simple systems. On the left are shown the relative positions of the ground states of the other nuclei with mass 12 which are unstable and decay by electron or positron emission to the ground state and lower levels of $^{12}$C.

A state may be regarded as long-lived if its lifetime is long compared with the nuclear transit time of $\sim 10^{-22}$ s, which is the time taken for a medium energy projectile to cross the nucleus.

The study of the spacing and properties of energy levels in nuclei is known as nuclear spectroscopy. The classification of the energy levels and an interpretation in terms of the internal structure of nuclei can be made using models, as was done in the case of atomic spectra and energy levels.

## NUCLEAR SPIN AND MAGNETIC MOMENTS

The total angular momentum of a nuclear state is given by $\sqrt{I(I + 1)}h/2\pi$. The quantum number $I$ is usually called the nuclear spin, although the total angular momentum may be composed of both spin and orbital angular momentum of the constituent nucleons. The $z$-component of the total angular momentum is $m_I(h/2\pi)$ where $m_I$ can take the $2I + 1$ values from $+ I$ to $- I$ in unit steps. It is found that the value of $I$ is related to the number of protons and neutrons as shown in Table 7.2. These values can be understood if we remember that the intrinsic spin quantum number of nucleon is $\frac{1}{2}$ and that the total number of nucleons is $A = Z + N$. Since the quantum numbers of orbital angular momentum always take integer values, the spin $I$ will be an integer or half-integer depending on whether $A$ is even or odd. The excited states of nuclei in general have different spins from the ground state, as can be seen from Fig. 7.6, but the rule embodied in Table 7.2 is obeyed for all states in the same nucleus.

The magnetic moment of a nucleus is given by

$$\mu_I = g_I\mu_N I, \tag{7.5}$$

where $g_I$ is the *nuclear g-factor* and $\mu_N$ is the *nuclear magneton*, which is smaller than the Bohr magneton by the ratio of the electron mass to the proton mass. The $z$-component of the magnetic moment is

$$\mu_{Iz} = g_I\mu_N m_I, \tag{7.6}$$

and the magnitude of the nuclear magnetic dipole moment is usually given in tables as the value of $\mu_{Iz}$ with $m_I = I$. This gives the magnetic moments of the proton and neutron as $+ 2.793\mu_N$ and $- 1.913\mu_N$, respectively. The magnetic moments are measured through processes involving interaction of the nuclear magnetic moment and an external field or interaction with the field due to the atomic electrons which leads to hyperfine splitting of the atomic spectral lines.

*Table 7.2.*  Relation between nuclear spin $I$ and nucleon number

| Z | N | I |
|------|------|--------------|
| Even | Even | Integer |
| Odd | Even | Half-integer |
| Even | Odd | Half-integer |
| Odd | Odd | Integer |

## NUCLEAR MAGNETIC RESONANCE

If nuclei possessing a magnetic moment are placed in a constant field $B_0$, the nuclei are distributed in a number of different energy states whose energies depend on the value of $m_I$. At equilibrium there are most nuclei in the lowest state and least in the highest state. However, if energy is supplied in the form of a quantum of electromagnetic radiation whose energy just matches the energy difference between one state and another, this energy can be absorbed by a nucleus which jumps to a higher state. The nucleus can subsequently fall back to a lower state with re-emission of a quantum of the same energy. Hence the process is one of resonant absorption and re-emission of radiation and is known as *nuclear magnetic resonance* (NMR).

The isotopes $^1$H, $^{13}$C, $^{19}$F and $^{31}$P all have $I = \frac{1}{2}$ and give the simplest pattern of only two energy states. The photon energy required to give excitation is

$$h\nu = (h/2\pi)\gamma B_0,$$

where $\gamma$ is the gyromagnetic ratio. For $^1$H, $\gamma/2\pi = 42.6$ MHz T$^{-1}$. Thus the resonant frequency is directly proportional to the applied field and for typical magnetic fields of 0.1–4.0 T the frequency is in MHz.

The amplitude of the NMR output signal decays, usually approximately exponentially with time, with a time constant $T_2$ which is known as the *transverse* or *spin–spin* relaxation time. Equilibrium magnetization is restored with a time constant $T_1$, which is known as the *longitudinal* or *spin–lattice* relaxation time. Both relaxation times represent an interaction between the nucleus and its surroundings, and hence NMR has an important role in studying the properties of molecules.

When a small linear magnetic field gradient is applied, the frequency of the radiation absorbed by each nucleus depends on its position, because frequency ∝ field ∝ position. The intensity of absorption depends on the number of nuclei present. Hence the absorption spectrum is a projection of nuclear spin density perpendicular to the magnetic field gradient. An image of nuclear spin density may then be reconstructed in a manner exactly analogous to X-ray computed tomography, which is described in Chapter 8. It is also possible to display separately the spatial distribution of the $T_1$ or $T_2$ relaxation time. At MHz frequencies, electromagnetic radiation can penetrate 10–100 cm of soft tissue, so that this method is suitable for medical imaging.

## ENERGY BALANCE IN NUCLEAR REACTIONS

In a nuclear reaction the initial nuclei may be transformed into different product nuclei with different masses, but the total energy must still be conserved. As an example we consider the reaction

$$a + A \rightarrow b + B,$$

where A is the target nucleus, assumed to be initially at rest and in its ground state, a is the projectile and B, b are the product nuclei. Applying conservation of energy gives

$$M_A c^2 + M_a c^2 + T_a = M_B c^2 + T_B + M_b c^2 + T_b,$$

where $T_a$, $M_a$ are the kinetic energy and rest mass of particle a, respectively, and similarly for the other particles. Hence

$$(M_A + M_a - M_B - M_b)c^2 = T_B + T_b - T_a \tag{7.7}$$

$$= Q. \tag{7.8}$$

The quantity $Q$ is called the *energy balance* or more usually the *Q-value* of the reaction. If $Q$ is positive, the reaction is said to be exothermic and the sum of the kinetic energies of the products is higher than the initial kinetic energy. If $Q$ is negative, the reaction is endothermic, and energy in the form of kinetic energy is "lost" or rather converted into mass; thus there is a *threshold* value for the projectile energy, and we must have $T_a \geq |Q|$, otherwise the reaction cannot happen.

If the nuclear masses can be taken as known from other sources, e.g. from isotopic masses measured with a mass spectrograph, measurement of the kinetic energies involved in the reaction can be used to verify Einstein's mass–energy relation. Such experiments have verified the relation to an accuracy of 1 part in $10^4$. On the other hand, if the validity of the relation is assumed, measurements of the $Q$-value can be used to determine the masses of unstable nuclei.

Equation (7.8) defines the least value of the $Q$-value corresponding to the reaction which leaves the product nucleus B in the ground state. If nucleus B is excited by an amount $E_{ex}$, the energy balance is

$$T_B + T_b - T_a = Q - E_{ex}.$$

We can apply these considerations to a variety of reactions and it is not necessary that there should be only two products. For example, the $Q$-value for the decay process

$$A \rightarrow x + X$$

is given by

$$Q = (M_A - m_x - M_X)c^2. \qquad (7.9)$$

If this $Q$-value is positive, the nucleus is unstable for this particular decay. If the $Q$-value is negative, the nucleus is stable against this decay and the magnitude of $Q$ represents the amount of energy required to remove the sub-group x from nucleus A leaving the product nucleus X in its ground state. This is known as the *separation energy* $S_{xA}$, so that

$$S_{xA} = (M_X + m_x - M_A)c^2. \qquad (7.10)$$

Some values of the separation energies in the nucleus $^{12}C$ are shown in Fig. 7.6.


## NUCLEAR BINDING ENERGIES

The nuclear binding energy is defined as the difference between the rest mass energy of the constituent nucleons and the rest mass energy of the nucleus, i.e.

$$\text{binding energy} = [ZM_p + (A - Z)M_n - M_A]c^2, \qquad (7.11)$$

where $M_p$, $M_n$, $M_A$ are, respectively, the rest masses of the proton, neutron and the nucleus with atomic number $Z$ and mass number $A$. This can also be written as

$$\text{binding energy} = [ZM_H + (A - Z)M_n - M_A']c^2,$$

where $M_H$ is the mass of the hydrogen atom and $M_A'$ is the mass of the neutral atom with atomic number $Z$ and mass number $A$. For all stable nuclei, the binding energy is positive and represents the amount of energy required to separate the nucleus into its constituent parts or, equivalently, the energy released when the constituents combine to form the stable nucleus. The magnitudes of the binding energies reach very high values ranging from 100 MeV for carbon to 1800 MeV for uranium.

The binding energy per nucleon is plotted in Fig. 7.7. It can be seen that, except for very light nuclei, the binding energy per nucleon is roughly constant, which implies that the addition of each nucleon adds the same amount of binding energy. On the other hand, the variation of this curve with mass number is of immense practical importance. For example, if a nucleus with mass number $\sim 240$ can be broken up into fragments which have mass numbers corresponding to the highest part of the binding energy/$A$ curve, the binding energy per nucleon is increased by $\sim 1$ MeV. Such an increase in binding energy would be accompanied by a release of

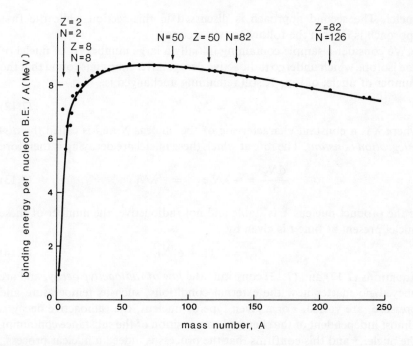

*Figure 7.7* The variation of the binding energy per nucleon with mass number $A$.

energy, mainly in the form of kinetic energy of the fragments. This method of releasing nuclear energy by breaking up a heavy nucleus is known as *fission* and is the basis for the operation of nuclear power reactors. Alternatively, if it is possible to combine two or more very light nuclei into one of heavier mass it is again possible to increase the binding energy per nucleon and release nuclear energy. This method of obtaining nuclear energy is called *fusion* and is the source of energy in stars.

## RADIOACTIVE DECAY

It is found that many of the heaviest nuclei are not completely stable but instead decay spontaneously to emit $\alpha$-particles (helium nuclei) and/or $\beta$-particles (electrons). There may also be $\gamma$-rays (photons) associated with the emission of $\alpha$- and $\beta$-particles. This process is known as *radioactive decay*, and can be studied in two ways. We can study an individual decay event and try to interpret it in terms of nuclear properties and nuclear forces, and we can also study the overall effect of the decay of a very large number of nuclei. The second approach is discussed in this section while the first approach is used in the following two sections.

We consider a sample containing initially a large number $N_0$ of nuclei of one isotope which undergo radioactive decay. After time $t$ it is found that the number of nuclei of type X still remaining unchanged is

$$N_X = N_0 e^{-\lambda t}, \tag{7.12}$$

where $\lambda$ is a constant characteristic of the nucleus X and is called the *disintegration constant*. The rate at which these nuclei are decaying is therefore

$$\frac{dN_X}{dt} = -\lambda N_0 e^{-\lambda t} = -\lambda N_X. \tag{7.13}$$

If the product nucleus Y is stable and not radioactive, the number of these nuclei present at time $t$ is given by

$$N_Y = N_0(1 - e^{-\lambda t}). \tag{7.14}$$

Equations (7.12) and (7.13) constitute the *law of radioactive decay*, and are obeyed no matter how the external conditions, such as temperature and pressure, are varied. For a given type of nucleus, the radioactive decay is almost independent of the chemical composition of the substance containing the nuclei,* and this confirms that the process is indeed a nuclear process.

It can be seen from eqn (7.13) that the number of nuclei which decay per unit time is $\lambda N$. (The minus sign in the equation simply indicates that the number is decreasing.) This is an average value, and careful measurements show that there is a statistical fluctuation about the mean value of $\lambda N$. In fact, the equations of radioactive decay can be deduced from probability arguments because we are dealing with a spontaneous and random process. We are not able to say which nucleus will disintegrate at what time, but only how many of a large group of nuclei will decay in a given interval. However, it is possible to determine a definite time in which a given fraction of the original nuclei will have decayed. We define the *half-life T* as the time taken for half the sample to decay, so that

$$N = \tfrac{1}{2}N_0 = N_0 e^{-\lambda T}$$

$$\therefore e^{\lambda T} = 2$$

$$T = \lambda^{-1} \log_e 2 = 0.693/\lambda. \tag{7.15}$$

---

* When a nucleus makes a radioactive transition emitting an $\alpha$-particle, electron or positron, the nuclear charge is changed and the parent atom becomes an ion. The parent atom also recoils. The effect of the change in the nuclear potential felt by the electron cloud is called "shaking". From all these causes the disruptive effect on a molecule containing a radioactive atom can be considerable. There can also be a very small change in the end-point energy in $\beta$-decay; very small chemical effects in nuclear $\beta$-decay have been predicted, but these can normally be ignored.

The values of $T$ measured for different radioactive nuclei vary over many orders of magnitude. The half-life of $^{238}U$ is $4.5 \times 10^9$ years while that of $^{214}Po$ is $1.6 \times 10^{-4}$ s. Some nuclei may decay by both $\alpha$-decay and $\beta$-decay. In this case, a disintegration constant can be measured for each type of decay and the total disintegration constant is $\lambda = \lambda_\alpha + \lambda_\beta$. This means that the half-life is $T = 0.693/(\lambda_\alpha + \lambda_\beta)$.

The emission of an $\alpha$-particle by a nucleus of mass number $A$ and atomic number $Z$ leads to another nucleus (often called the daughter nucleus) with mass number $A - 4$ and atomic number $Z - 2$, whereas the emission of an electron leads to a daughter nucleus with the same mass number and atomic number $Z - 1$. (We are assuming here the conservation of charge and the conservation of the total number of nucleons.) In either case, the daughter nucleus is itself likely to be radioactive and to decay into yet another nucleus. This successive decay process gives rise to a *radioactive chain* which terminates when a stable nucleus is reached. If the first nucleus has a very long half-life compared with the other nuclei in the chain and the system is undisturbed so that no decay products can escape, it is possible for the system to reach *secular equilibrium* in which the numbers of nuclei of each type remain constant. The numbers of nuclei of each type then satisfy the condition

$$\lambda_1 N_1 = \lambda_2 N_2 = \lambda_3 N_3 = \ldots .$$

This condition is useful in the study of uranium-bearing minerals because of the long half-life of uranium.

## DERIVATION OF RADIOACTIVE DECAY LAWS

It is assumed that disintegration is subject to the laws of probability, is independent of the past history of the nucleus, and is the same for all nuclei of the same type. Then the probability of decay $p$ is proportional to the length of the time interval $\Delta t$, i.e.

$$p = \lambda \, \Delta t,$$

where $\lambda$ is the distingeration constant. The probability that the nucleus will not decay is

$$1 - p = 1 - \lambda \, \Delta t.$$

If we start with $N_0$ atoms, the number left after time $\Delta t$ is

$$N_1 = N_0(1 - \lambda \, \Delta t),$$

and after another interval $\Delta t$ is

$$N_2 = N_1(1 - \lambda \, \Delta t) = N_0(1 - \lambda \, \Delta t)^2.$$

After $n$ time intervals of length $\Delta t$

$$N = N_0(1 - \lambda \, \Delta t)^n,$$

but total time $t = n \, \Delta t$

$$\therefore N = N_0\left(1 - \lambda \frac{t}{n}\right)^n.$$

Because

$$e^{-x} = \lim_{n \to \infty} \left(1 - \frac{x}{n}\right)^n$$

we have

$$N = N_0 e^{-\lambda t},$$

which is eqn (7.12). Thus the law of radioactive decay is a statistical law and is result of a large number of events subject to laws of probability. It follows from eqn (7.12) that

$$\frac{dN}{dt} = -\lambda N,$$

and this leads to definition of *activity* as

$$A = \lambda N.$$

Experimentally, we measure the quantity $A' = c\lambda N$, where $c$ is the detector efficiency, but since

$$\frac{A(t_1)}{A(t_2)} = \frac{N(t_1)}{N(t_2)}$$

$c$ cancels out. From eqn (7.13) we have

$$A = \lambda N_0 e^{-\lambda t}, \quad \log A = \log \lambda N_0 - \lambda t,$$

so that a plot of $\log A$ against $t$ is a straight line.

## RADIOACTIVE CHAINS

A radioactive sequence of the form

$$1 \to 2 \to 3 \to \ldots \to \text{stable nucleus } s$$

is described by the equations

$$\frac{dN_1}{dt} = -\lambda_1 N_1$$

$$\frac{dN_2}{dt} = +\lambda_1 N_1 - \lambda_2 N_2$$

$$\frac{dN_3}{dt} = +\lambda_2 N_2 - \lambda_3 N_3$$

$$\frac{dN_s}{dt} = +\lambda_{s-1} N_{s-1}.$$

These are called the Bateman equations, and can be solved by standard methods.

It is also possible for the radioactive chain to branch, thus

$$1 \to 2 \to 3 \begin{smallmatrix} \nearrow & 4 & \searrow \\ & & 6 \to \dots \\ \searrow & 5 & \nearrow \end{smallmatrix}$$

If the decay $3 \to 4$ goes by $\alpha$-decay and the decay $3 \to 5$ goes by $\beta$-decay, then the decay $4 \to 6$ will go by $\beta$-decay and the decay $5 \to 6$ will go by $\alpha$-decay.

## UNITS OF RADIOACTIVITY

The original unit of radioactivity was the *curie* (Ci). An isotope has an activity of 1 Ci when its rate of decay is $3.7 \times 10^{10}$ disintegrations per second. This is roughly equal to the number of disintegrations per second in 1 g of pure radium. The SI unit of activity is the *becquerel* (Bq) which is equal to one disintegration per second.

The becquerel is an absurdly small unit for modern usage. Many ordinary substances are slightly radioactive: for example, tea leaves have an activity of about 830 $Bq.kg^{-1}$ and instant coffee has an activity of about 1640 $Bq.kg^{-1}$, depending on the country of origin and method of preparation. These values fall within the range of activity of low level waste from hospitals, industry and nuclear power plants, which is disposed of by burial in near-surface sites. In diagnostic imaging, the activity of the radio pharmaceutical injected into the patient is typically about 10 mCi or 370 MBq.

As we have already seen in Chapters 2, 4 and 5 the passage of charged particles or photons through matter gives rise to ionization. The various types of charged particles are referred to as directly ionizing radiations, whereas photons are indirectly ionizing radiations, as they may interact with matter by one of several processes from which charged particles result. A charged particle loses energy along its track by repeated interactions with the absorbing medium and the energy absorbed by the medium is known as the *absorbed dose*. The original unit of dose, the rad, corresponds to an absorption of $10^{-2}$ J kg$^{-1}$ by matter in the place of interest. The SI unit of absorbed dose is the *gray* (Gy) which is equal to absorption of 1 J kg$^{-1}$. In terms of energy absorbed, the rad is a small unit and a dose of $10^6$ rads will raise the temperature of water by only 2.4 K. However, the absorption of energy is very non-uniform on an atomic scale, as it takes place along the tracks of the charged particles. For this reason, radiation can be damaging even when the dose is too low to heat the material in bulk. This is particularly true for living organisms, and a whole body dose of 5 Gy may be lethal to man. The energy absorbed per unit length of track varies for different charged particles, so that equal doses will not necessarily give rise to the same biological effects. A weighting factor, known as the quality factor (QF) allows for this. The dose multiplied by the QF is known as the *dose equivalent* (H). It can be seen from Table 7.3 that, dose for dose, neutrons and $\alpha$-particles are more damaging than X-rays, $\gamma$-rays and electrons.

The original unit for dose equivalent was the rem. The SI unit is the *sievert* (Sv). One millisievert (mSv) is roughly equivalent to the dose delivered in 50 chest X-rays or four times the average annual dose due to medical examination. The average annual dose equivalent for ordinary members of the UK population is listed in Table 7.4, using 1984 figures for the UK. Among the natural contributions there are natural radionuclides present in the body, e.g. $^{14}$C and $^{40}$K, and inhaled decay products from radon and thoron gases emanating from the Earth's surface, especially in granite regions, and from building materials. About 90% of the dose due to medical practices is due to diagnostic radiology. The miscellaneous artificial dose includes extra exposure to cosmic rays due to air travel, use of radioactive materials in consumer products and natural radioactivity in fly ash released during the burning of coal. The contribution from coal burning, mainly in power

*Table 7.3.* The quality factor (QF) for various types of radiation

| Radiation | QF |
|---|---|
| X-rays, $\gamma$-rays | 1 |
| $\beta$-particles (electrons) | 1 |
| $\alpha$-particles | 10 |
| Thermal neutrons | 2–5 |
| Fast neutrons | 5–10 |

Table 7.4. The average annual dose equivalent to a member of the UK population in 1984

| Source | Dose equivalent (mSv) |
|---|---|
| *Natural* | |
| Cosmic rays | 0.300 |
| Terrestrial $\gamma$-rays | 0.400 |
| Internal radiation | 0.370 |
| Radon gas | 0.700 |
| Thoron gas | 0.100 |
| | 1.870 |
| *Artificial* | |
| Medical | 0.250 |
| Miscellaneous | 0.011 |
| Fallout | 0.010 |
| Occupational exposure | 0.009 |
| Radioactive waste disposal | 0.002 |
| | 0.282 |
| *Total* | 2.152 |

stations, is slightly greater than the contribution from radioactive waste disposal.

For X-rays and $\gamma$-radiation, it is sometimes convenient to measure the quantity of radiation or intensity of the beam without reference to a particular medium which is to be irradiated. A difficulty with this procedure is that if there is no medium there is no ionization effect to measure. It is the convention, therefore, to measure the ionization produced per kilogram of air at standard temperature and pressure. This quantity is known as the *exposure* and was measured by a unit called the roentgen (R) where 1 R = $2.58 \times 10^{-4}$ C kg$^{-1}$. In certain circumstances, an exposure of 1 R will deliver a dose of approximately 1 rad. However, this is not generally true and the two units are different. The SI unit is C kg$^{-1}$, but it has no special name.

## ALPHA DECAY

We now examine a single decay process leading to the emission of an $\alpha$-particle, which can be represented as

$$^{A}_{Z}X \rightarrow {}^{4}_{2}He + {}^{A-4}_{Z-2}Y. \qquad (7.16)$$

The production of the daughter nucleus Y can be verified by chemical means, and the $\alpha$-particle can be identified as a helium nucleus by chemical and physical means, such as measurement of $q/m$ or identification of the optical spectra emitted after the $\alpha$-particle has slowed down and picked up two electrons. Measurements of the kinetic energies of the emitted $\alpha$-particles show that these are of the order of a few MeV, which again confirms that the decay process must be a nuclear transformation.

The $Q$-value for the decay process represented by eqn (7.16) is

$$Q_\alpha = \{M_X - M_\alpha - M_Y\}c^2. \tag{7.17}$$

For heavy nuclei this $Q$-value is positive so that the process can indeed occur spontaneously. The kinetic energy $T_\alpha$ of the $\alpha$-particle can be calculated from the $Q$-value using the conservation laws for energy and momentum. Assuming that the nucleus X is at rest when it decays and that the kinetic energies can be treated non-relativistically, we have

$$0 = M_\alpha v_\alpha - M_Y v_Y$$

$$Q_\alpha = \tfrac{1}{2}M_\alpha v_\alpha^2 + \tfrac{1}{2}M_Y v_Y^2,$$

and hence

$$Q_\alpha = \tfrac{1}{2}M_\alpha v_\alpha^2\left(1 + \frac{M_\alpha}{M_Y}\right),$$

or

$$T_\alpha = \tfrac{1}{2}M_\alpha v^2 = Q_\alpha\bigg/\left(1 + \frac{M_\alpha}{M_Y}\right). \tag{7.18}$$

Measurements of the kinetic energies of $\alpha$-particles emitted in decays from many nuclei show that the maximum value is always in agreement with eqn (7.18). However, it is observed that for a given decay there is a discrete spectrum of $\alpha$-particle energies, with groups of $\alpha$-particles having energies lower than that predicted by eqn (7.18). This result can be understood by reference to Fig. 7.8. We have already noted that nuclei have discrete spectra of excited states, so that that $\alpha$-decay process must always leave nucleus Y in its ground state or one of the excited states. Thus, the kinetic energy of the $\alpha$-particle must take discrete values corresponding to the energy difference between the ground state of X and a definite value in Y. The missing energy appears in the form of a $\gamma$-ray, when nucleus Y falls from an excited state to its ground state through an electromagnetic transition, and this explains why $\gamma$-rays are observed in association with $\alpha$-decay. The energy of the $\gamma$-ray is therefore given by

$$E_\gamma = Q_\alpha - T_\alpha(1 + M_\alpha/M_Y). \tag{7.19}$$

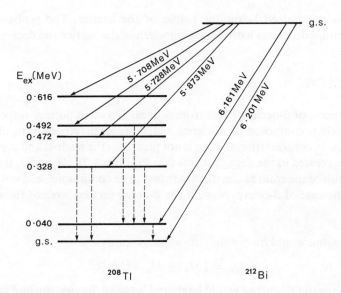

*Figure 7.8.* The $\alpha$-decay of the ground state of $^{212}$Bi. The unbroken arrows represent transitions associated with the emission of an $\alpha$-particle and the dashed arrows indicate emission of $\gamma$-rays.

Another very important feature of the $\alpha$-decay process may be examined using Fig. 7.5b. Previously, we used Fig. 7.5a to note that the Coulomb interaction acted as a barrier to keep charged projectiles out of the nucleus. However, if we now think of the $\alpha$-particle being formed inside the nucleus the barrier acts to keep the $\alpha$-particle inside the nucleus even though the energy of the $\alpha$-particle is above the zero of energy. Thus, we may take the probability of decay per unit time, which is the same thing as the dis-integration constant, as the product of the probability that four nucleons resembling an $\alpha$-particle appear at the surface of the nucleus near to the barrier multiplied by the probability for penetration through the barrier. Once the $\alpha$-particle gets through the barrier it can escape from the nucleus because of its kinetic energy and the Coulomb repulsion. If we take a classical view of the problem, the probability for penetration is zero unless $T_\alpha$ is greater than the height of the barrier, but a quantum mechanical calcula-tion using Schrödinger's equation shows that there is a finite probability for the probability density to leak through the barrier and for free $\alpha$-particles to appear on the outside. This is known as the *tunnel effect*. The height of the barrier can be determined from Rutherford scattering from the nucleus, and comparison with $\alpha$-decay shows that the kinetic energy of the emitted

$\alpha$-particle is indeed below the height of the barrier. The probability for penetration decreases with increasing width of the barrier and decreasing $T_\alpha$.

## BETA DECAY

The process of $\beta$-decay differs from $\alpha$-decay in two important respects. The $\alpha$-particle is composed of nucleons which are already present in the initial nucleus. In contrast, the electron is not present in the nucleus and must therefore be created in the decay process (see Problem 7.7). Secondly, the energy spectrum of the emitted electrons is found to be continuous, and not discrete as in the case of $\beta$-decay. Now, if the $\beta$-decay process were of the form

$$_Z^A X \rightarrow e^- + {}_{Z+1}^A Y,$$

the $Q$-value would have a definite value given by

$$Q_\beta = \{M_X - M_\alpha - M_Y\}c^2,$$

and the available energy would be shared between the electron and nucleus Y in a very definite manner determined by the conservation laws. In fact, the maximum electron energy observed is in agreement with this prediction, but electrons emitted in decays of other nuclei of the same type are observed to have energies between this maximum value and zero, and thus the energy spectrum is continuous. The average energy is considerably less than the maximum energy. Also, if the decaying nuclei are placed in a calorimeter, with thick walls which stop the electrons, the heat produced is related to the average energy and not to the maximum energy. If a single decay process is observed by recording the ionization caused by the recoiling nucleus Y and by the emitted electron it is found that energy and momentum are apparently not conserved.

An explanation of the $\beta$-decay process was first proposed by Pauli and developed by Fermi into a consistent theory of $\beta$-decay. It is assumed that the basic process of $\beta$-decay is the decay of a neutron

$$n \rightarrow p + e^- + \bar{\nu}, \tag{7.20}$$

where $\bar{\nu}$ represents the antiparticle of a new particle called the *neutrino*. The nuclear decay is now

$$_Z^A X \rightarrow {}_{Z+1}^A Y + e^- + \bar{\nu}. \tag{7.21}$$

(The neutrino is denoted by $\nu$. The difference between the neutrino and its antiparticle will not be considered here. We also ignore the difference between the neutrino associated with an electron $\nu_e$ and the neutrino associated with a muon $\nu_\mu$.) Because the energy can be shared between the three particles in the final state, the energy of the electron no longer has a unique value.

The properties of the neutrino are as follows:

(*i*)   It has no charge. This follows from eqn (7.20) and conservation of charge.

(*ii*)   It has a spin of $\frac{1}{2}(h/2\pi)$. This follows from eqn (7.20) and conservation of angular momentum.

(*iii*)   It has an extremely small or zero rest mass. This follows because the end-point of the energy spectrum of the electrons goes up to the maximum value allowed without loss of any energy required to create a neutrino with a sizeable rest mass. The rest mass is usually taken to be identically zero.

(*iv*)   It interacts very weakly with matter. If this were not so, the neutrinos would have been stopped in the calorimeter experiment and their energy would have been absorbed by the calorimeter, giving a temperature rise corresponding to the maximum available energy.

Because the neutrino interacts very weakly it is difficult to detect its presence. However, independent evidence for the existence of the neutrino comes from processes produced by the rather intense neutrino flux emitted from nuclear reactors due to neutron decay. In particular, the reaction

$$p + \bar{\nu} \rightarrow n + e^{+}$$

has been studied. The presence of the positron is detected through the $\gamma$-rays emitted when it undergoes pair annihilation, and the neutron is detected through emission of $\gamma$-rays when it is captured in a suitable heavy nucleus.

The process of $\beta$-decay can leave the daughter nucleus Y in an excited state, as in the case of $\alpha$-decay. For each excited state there is a new end-point given by $Q_\beta - E_{ex}$, and the electrons corresponding to excitation of this state can take any energy from zero up to the end-point. Thus, the total spectrum may be a very complicated sum of several continuous spectra.

Many unstable light nuclei decay by positron emission. The basic process in this case is

$$p \rightarrow n + e^{+} + \nu,$$

which can be described by the same basic theory as electron emission.

## FISSION

The fission process can be induced in heavy nuclei by supplying energy in the form of a $\gamma$-ray or kinetic energy carried by a projectile. Neutron-induced fission can be represented by the equation

$$n + {}^{235}_{92}U \rightarrow {}^{236}_{92}U^* \rightarrow X + Y.$$

The products X and Y are known as the *primary fission fragments*. The most probable mass numbers for these nuclei are $A \approx 96, 140$, respectively; symmetric fission is much less probable. Because of the relation between $Z$ and $N$ for stable nuclei (see Fig. 7.1), these fission fragments have too many neutrons and are not stable. They can emit neutrons — 99% of these are *prompt* neutrons emitted $\sim 10^{-14}$ s after fission, and the remaining 1% are *delayed* neutrons emitted a few seconds or minutes after fission. The fragments can also $\beta$-decay to change a neutron into a proton. This leads to a chain of decays until a stable nucleus is reached, e.g.

$$^{140}Xe \xrightarrow{\beta^-} {}^{140}Cs \xrightarrow{\beta^-} {}^{140}Ba \xrightarrow{\beta^-} {}^{140}La \xrightarrow{\beta^-} {}^{140}Ce \text{ (stable)}.$$

Similar processes can lead to radioactive isotopes of iodine and strontium, which are dangerous to man.

The energy released in the fission process appears mainly as kinetic energy of the fission fragments. Fission of one $^{235}U$ nucleus yields (average values in MeV):

| | |
|---|---:|
| kinetic energy of fission fragments: | 165 |
| kinetic energy of fission neutrons: | 5 |
| $\beta^-$ energy: | 5 |
| $\gamma$-ray energy: | 7 |
| neutrino energy, delayed $\beta$-emission, and $\gamma$-emission from radioactive decay: | 20 |
| | 202 MeV |

Each fission process leads to the release of several neutrons. If these neutrons in turn cause fission the process becomes a self-sustaining *chain reaction*. In practice, this is difficult to achieve. Natural uranium contains 0.718% $^{235}U$, 0.006% $^{234}U$ and 99.27% $^{238}U$. When bombarded with neutrons both $^{235}U$ and $^{238}U$ can undergo fission or neutron capture $U(n, \gamma)$. For $^{235}U$, the cross-sections for both processes increase with decreasing neutron velocity and at thermal neutron energies ($\sim 0.025$ eV) about 85% of the reactions give fission. Energies of the emitted neutrons cover a range up to 10 MeV, with the most probable energy a little less than 1 MeV. For $^{238}U$, fission does not occur for neutron energies below 1 MeV, and the capture cross-section shows the same general behaviour for $^{235}U$, but for certain neutron energies between 5 and 500 eV the capture cross-section can be $10^3$ times higher than normal. This means that it is impossible to have a self-sustaining reaction in natural uranium since the neutrons emitted by fission of $^{235}U$ have, in general, too little energy to give fission in $^{238}$U but are captured by $^{238}$U instead.

The difficulty can be overcome by enriching the uranium sample to contain a much higher proportion of $^{235}$U. This procedure is used in *fast* reactors. Hence, techniques of isotope separation have become very important. In *thermal* reactors, the uranium fuel is mixed with a *moderator* containing nuclei of light mass so that the neutrons make collisions and lose kinetic energy until they come down to thermal energies.

Neutron capture in $^{238}$U can be used to produce $^{239}$Pu, which is a nuclear fuel very similar to $^{235}$U. If $^{239}$Pu is used in a fast reactor and enclosed with uranium, the capture of escaping neutrons by the uranium leads to more plutonium. This process is the basis of the *breeder* reactor.

## FUSION

Fusion reactions are responsible for the large amount of energy from stars. The basic process is the fusion of four protons into a helium nucleus and two positrons. For controlled energy release on earth, the fusion of a deuteron and a triton into helium and a neutron seems to be the most promising process. The reaction is

$$^{2}_{1}\text{H} + {}^{3}_{1}\text{H} \rightarrow {}^{4}_{2}\text{He} + \text{n} + 17.6 \text{ MeV},$$

and the fuel, the isotopes of hydrogen, is available in an almost inexhaustible supply from the sea.

Particle energies of approximately 10 keV are required to overcome the Coulomb barrier between them and these energies correspond to a temperature of $\sim 10^{8}$ K. At such a temperature, atoms are dissociated into a *plasma* of nuclei and electrons. Much of fusion research is concerned with containment of such a plasma. An interesting alternative possibility involves the catalysis of *cold* fusion by muons, by formation of molecules of the form (dt$\mu$). The small size of the muon orbit helps to pull the deuteron and triton closer together.

## THE NUCLEAR FORCE

Information about the nuclear force is built up by studying the scattering of nucleons from nucleons and the behaviour of the simplest bound systems of two or three nucleons, and also by studying the properties of complex nuclei.

An important result can be deduced from the observation that the binding energy per nucleon is approximately constant, which means that the total binding energy is proportional to the mass number $A$. Now, if every nucleon in the nucleus interacts with every other nucleon the binding energy would be approximately proportional to the number of pairs, which is $\frac{1}{2}A(A-1)$, so that in a large nucleus the binding energy would be proportional to $A^2$. This is in disagreement with the experimental observation, which suggests instead that each nucleon interacts with a limited number of nearest neighbours. When another nucleon is added to the nucleus it interacts with the same number of nearest neighbours and so adds a constant amount to the binding energy and to the volume. This phenomenon is known as *saturation*, and is also a characteristic of intermolecular forces in fluids. A diagram of the intermolecular potential is shown in Fig. 7.9. The attractive part is due to the so-called van der Waals force, and the short-range repulsive part is due to the repulsion of overlapping electron shells. It is found that the nucleon–nucleon interaction can be represented by a potential of similar shape, although the magnitudes and distances are quite different. In the nuclear case, the radius of the repulsive part is $\sim 0.4$ fm and the range of the attractive part is $\sim 1.4$ fm. The overall effect of the nuclear force must be attractive, otherwise it could not lead to bound systems, but a completely attractive short-range force would lead to a collapse of the nucleus into a volume much smaller than the observed value. The effect of the exclusion principle also assists in producing saturation.

Further experimental evidence shows that the nuclear force is *spin-*

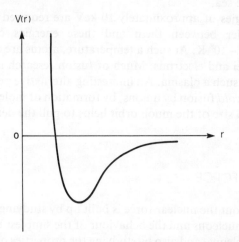

*Figure 7.9.* A typical form for the intermolecular potential. The nucleon–nucleon potential can be represented by a similar function.

*dependent*, i.e. it depends on the relative spin of the two interacting nucleons. To an accuracy of a few per cent, at least, the force is found to be *charge-independent*, i.e. the same between a pair of protons, a pair of neutrons, or a proton and a neutron. For interactions between protons, however, we must add the repulsive Coulomb force to the nuclear force. In general, the effect of the electromagnetic interaction is weaker than the nuclear interaction by a factor of 1/137, although the Coulomb repulsion between a large number of protons in a heavy nucleus can overcome the attractive nuclear force and lead to instability.

The process of $\beta$-decay occurs through the *weak interaction*, which is different in form and effect from the strong interaction and weaker by a factor of $10^{-13}$. It is the only interaction experienced by the neutrino.*

## NUCLEAR MODELS

As we have noted before, the classification and interpretation of the properties of complex nuclei proceeds quite naturally through the use of the appropriate models.

Many of the gross properties of nuclei, such as the increase of binding energy and volume with mass number, and the property of saturation, indicate that it would be profitable to use a model which relates the properties of the nucleus to those of a drop of incompressible fluid. In this *liquid drop model* we do not attempt to describe the motion of individual nucleons but instead concentrate on the collective behaviour due to the coherent motion of all the nucleons. An example of the use of this model arises in the derivation of the *semi-empirical mass formula*. The nuclear binding energy has a leading term which is proportional to mass number. This is the volume term

$$E_v = a_v A,$$

where $a_v$ is some constant to be determined empirically. Now, using the liquid drop model we expect that the binding energy should be reduced by a surface term due to the fact that the nucleons on the surface do not have a full compliment of nearest neighbours. This is proportional to the surface area, so that

$$E_s = -a_s A^{2/3}.$$

* We ignore the gravitational interaction experienced by all particles since this has a relative strength of $10^{-38}$.

We must also subtract a term representing the Coulomb repulsion of the protons which is inversely proportional to the radius

$$E_C = - a_C Z(Z - 1)/A^{1/3},$$

and also a symmetry term

$$E_r = - a_r(A - 2Z)^2/A,$$

which arises from the fact that in the absence of Coulomb repulsion the nuclei with $Z = N$ are most stable. Finally, there is a pairing energy $E_\delta$ which is positive for even–even nuclei, negative for odd–odd nuclei and zero for the rest. Hence the formula for the nuclear binding energy is

$$\text{binding energy} = E_v + E_s + E_C + E_r + E_\delta,$$

and the mass formula is

$$M = ZM_p + (A - Z)M_n - \text{binding energy}/c^2.$$

These formulae give a good overall description of the binding energies and masses of nuclei but do not describe certain specific features.

The existence of non-zero quadrupole moments shows that many nuclei are non-spherical. This phenomenon can be described rather well within the framework of the liquid drop model by allowing the drop to be deformed into a non-spherical shape. In order to have a complete model appropriate for the description of nuclei it is necessary to impose the condition that the excited states of the system are quantized. In the liquid drop model, this leads to modes of excitation corresponding to vibrations of the surface and rotations of the whole system. Such vibrational and rotational states are observed in a number of nuclei.

A somewhat similar approach is taken in the *compound nucleus model* of nuclear reactions. In this case it is assumed that when a low-energy nucleon impinges on the target nucleus it is absorbed and shares its kinetic energy rapidly and collectively with the nucleons in the nucleus, so forming the compound nucleus of $A + 1$ nucleons. This compound nucleus must be in an excited state owing to the excess energy brought in by the projectile, and will eventually decay by emitting a particle or $\gamma$-ray. Because the energy of the projectile is shared randomly among all the nucleons, the state of the compound nucleus so formed is relatively long-lived, and hence from eqn (7.4) its width is small. For a given process, a plot of the total cross-section as a function of energy, such as that shown in Fig. 7.10, shows sharp peaks known as *resonances* whenever the incident energy coincides with the excitation energy of one of the states of the compound nucleus. The widths of the resonances may be as narrow as eV corresponding to lifetimes of $\sim 10^{-14}$ to $10^{-15}$ s.

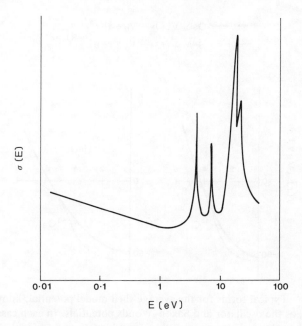

*Figure 7.10.* The behaviour of the total cross-section as a function of projectile energy showing the presence of sharp resonances.

Despite the usefulness of these macroscopic models, we would also like to be able to describe the motion of individual nucleons inside the nucleus. There is a substantial body of evidence which indicates that the nucleons exist in well-defined quantized states and that the nucleons form shells in a manner similar to electrons in atoms. For example, it is found that when the proton number $Z$ or the neutron number $N$ is equal to one of the *magic numbers* 2, 8, 20, 28, 50, 82 or 126 the corresponding nuclei possess exceptional stability. The binding energy per nucleon is not such a smooth function as Fig. 7.7 may suggest, and there are sharp rises in binding energy/$A$ at the magic numbers. It can be seen from Fig. 7.3 that the quadrupole moment goes to zero at these numbers. There are many other indications that the magic numbers confer stability and, by analogy with the properties of atoms, this is interpreted as evidence for shell closure. Studies of nuclear reactions at medium and high energies show that when a single nucleon is added to or removed from a nucleus, its behaviour in the nucleus can be described by a single-particle wavefunction characterized by a set of quantum numbers for that nucleon. This implies that the nucleon maintains to a considerable extent its independent motion in the nucleus.

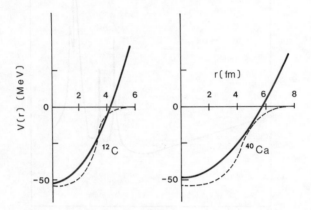

*Figure 7.11.* Typical forms for the average shell model potential, known, respectively, as the oscillator and Saxon–Woods potentials. In each case the parameters have been chosen to give the correct binding energy for the least bound nucleon.

These observations have led to the development of the nuclear *shell model* in which it is assumed that each nucleon moves independently in a potential which represents the averaged effect of its interaction with the other nucleons. Typical forms of this potential are shown in Fig. 7.11; in the centre of the nucleus each nucleon is surrounded by the same number of nearest neighbours so that the potential is constant but in the surface that number of neighbours is less and so the potential tends to zero. The wavefunctions and energies of the individual nucleons can be determined by solving the Schrödinger equation for this potential, exactly as in the case of one-electron atoms. In order to obtain shell closure at the observed magic numbers it is necessary to include a spin–orbit splitting, which is of the same form as that observed in atoms but is stronger and of opposite sign. For protons we must also include the Coulomb potential so that the individual energies of protons are slightly different from those of neutrons. Energy levels of a single nucleon in a realistic shell model potential are shown in Fig. 7.12. By assuming that the nucleon states are filled up in order of increasing principal quantum number and increasing energy (decreasing binding energy) we may determine the nucleon configuration in each nucleus, just as we determined the electronic configuration in atoms, and the nucleons fill the available

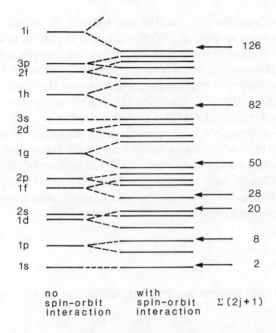

*Figure 7.12.* Single-particle energy levels in a realistic shell model potential.

energy levels up to the Fermi energy, as in the case of the free electron theory of metals. With this simple form of the shell model it is possible to predict the properties, such as the nuclear spin and magnetic moment, of nuclei which have one nucleon outside a closed shell. When there are several nucleons in an incomplete shell it is necessary to take account of the part of nucleon–nucleon interaction not included in the average potential, but the fact that the effect of this *residual interaction* is generally weak makes it possible to consider it as a correction which modifies but does not destroy the independent motion. When there are only a few valence nucleons in the unfilled shell, the residual interaction has a pairing effect which gives rise to the pairing energy $E_\delta$ in the mass formula and causes pairs of like particles to couple to spin zero, but for a larger number of valence nucleons this gives way to collective behaviour and deformation of the nucleus.

A similar approach is taken in the *optical model* for medium and high-energy nuclear scattering in which it is assumed that the interaction between the projectile and the nucleus can be represented by an average potential. It is the parameters of this potential which are normally determined by analyses of elastic scattering data.

It is an encouraging feature of the development of nuclear models that the shell model with residual interaction can indicate the origin of collective effects. It may seem surprising, at first sight, that the nucleons can maintain any independent motion when they are interacting through a strong short-range force. However, when a pair of nucleons collide or interact they change their energy and momentum, which means that they change their quantum numbers. But if this collision takes place inside the nucleus all the states up to the Fermi level are filled so that scattering into them is forbidden by the exclusion principle, and hence the only allowed collisions are relatively energetic ones which take a nucleon above the Fermi energy. Thus, the exclusion principle acts to reduce drastically the number of collisions within the nucleus. Clearly, this restriction is least effective for those nucleons nearest to the Fermi energy so that the residual interaction is most important for the most energetic nucleons in the unfilled shell.

## PROBLEMS

**7.1.**  The uniform nuclear density distribution is given by

$$\rho(r) = \rho_0, \ r \leqslant R;$$

$$\rho(r) = 0, \ r > R.$$

Show that the mean square radius is $3R^2/5$.

**7.2.**  The equivalent radius $R$ increases as $1.2A^{1/3}$, approximately. Use this value to calculate a value for the nuclear density. Explain why this value is different from the value obtained in Chapter 4.

*Solution.*  $\rho = 0.23 \times 10^{18} \ \mathrm{kg \ m^{-3}}$

**\*7.3.**  Consider the ground state of a one-electron atom in which the electron is replaced by a negative muon. Show that the ratio of the rms radii for the electron and muon orbits is equal to the inverse ratio of the masses of the particles.

**7.4.**  Using eqn (4.3) deduce the differential cross-section for low energy $\alpha$-particle scattering.

*Solution.*  Equation (4.3) gives the number of $\alpha$-particles $N(\theta)$ scattered through an angle between $\theta$ and $\theta + d\theta$ into a solid angle $2\pi \sin \theta \ d\theta$, for an incident number of $\alpha$-particles $N_0$ and $ns$ nuclei per unit area of target. Hence

$$d\sigma = N(\theta)/N_0 \ ns$$

and

$$\frac{d\sigma}{d\Omega} = \frac{Z^2 e^4}{(4\pi\epsilon_0)^2 4 T_\alpha^2} \frac{1}{\sin^4 \frac{1}{2}\theta}.$$

**\*7.5.**  The deuteron d is the isotope $^2$H of hydrogen so that the result of the reaction

$$d + A \rightarrow p + B$$

is to add a neutron to nucleus A. Show that the separation energy of this neutron from nucleus B is equal to the sum of the Q-value for the (d, p) reaction and the binding energy of the deuteron.

**7.6.**  Find the mean lifetime of a radioactive nucleus whose half-life is 10 years.

*Solution.*  Mean lifetime = $T_{\frac{1}{2}}/0.693 = 14.4$ years

**7.7.**  Use the uncertainty principle to show that the electron cannot be a constituent of the nucleus.

*Solution.*  According to the uncertainty principle $\Delta x \, \Delta p \approx h$, hence if the electron is confined in a nucleus of radius $R$ it gains momentum

$$\Delta p \approx \frac{h}{2R} \approx \frac{6.6 \times 10^{-34}}{2 \times 10^{-14}} = 3.3 \times 10^{-20} \text{ kg m s}^{-1}$$

The energy of the electron is given by

$$E^2 = \Delta p^2 c^2 + m_0^2 c^4,$$

and for the electron $m_0 c^2 = 0.51$ MeV. In this case we have

$$\Delta p \, c = \frac{3.3 \times 10^{-20} \times 3 \times 10^8}{1.6 \times 10^{-13}} = 62 \text{ MeV},$$

so that the rest mass energy is negligible and the electron has a kinetic energy of the order of 60 MeV.

The Coulomb attraction of the protons is not strong enough to hold such an energetic electron in the nucleus.

**7.8.**  Calculate the energies of the $\gamma$-rays emitted in the transition represented by dashed arrows in Fig. 7.8 and check that they obey eqn (7.19).

**7.9.** Show that the $Q$-value for the reaction

$$a + A \rightarrow a + x + X$$

is equal to $-S_{xA}$.

**7.10.** An $\alpha$-particle is elastically scattered from a proton which is initially at rest. Show that the maximum possible scattering angle for the $\alpha$-particle is $14° \, 30'$.

*Solution.* The conservation laws give

$$T_0 = T_1 + T_2$$

$$p_0 = p_1 \cos \theta_\alpha + p_2 \cos \theta_p$$

$$0 = p_1 \sin \theta_\alpha - p_2 \sin \theta_p$$

where $T_0, p_0$ and $T_1, p_1$ are, respectively, the initial and final energies and momenta of the $\alpha$-particle and $T_2, p_2$ are the final energy and momentum for the proton.

Eliminating $\theta_p$ and $p_2$ yields

$$\left(1 - \frac{M_p}{M_\alpha}\right)p_0^2 - 2p_0 p_1 \cos \theta_\alpha + \left(1 + \frac{M_p}{M_\alpha}\right)p_1^2 = 0.$$

This equation has a solution only when $\cos^2 \theta_\alpha \geq 15/16$.

**7.11.** (*a*) Find the electrostatic force between two protons separated by a distance of 1 fm and also the gravitational force between them. The gravitational constant is

$$G = 6.67 \times 10^{-11} \text{ N m}^2 \text{ kg}^{-2}.$$

(*b*) If the nuclear potential energy between two protons separated by a distance $r$ is given by

$$V(r) = \frac{q^2}{r} \exp\left(-\frac{r}{R}\right)$$

where $2\pi \, q^2/hc = 15$ and $R = 1.4$ fm, find the force between two protons separated by a distance of 1 fm.

*Solution.* (*a*) $F_E = 2.30 \times 10^2$ N, $F_G = 1.86 \times 10^{-34}$ N.
(*b*) $F_N = 3.98 \times 10^5$ N.

# 8 | Instrumentation and Applications

The purpose of this chapter is to outline some practical developments of the ideas we have introduced in earlier chapters, and to describe some applications in a variety of fields.

## PARTICLE ACCELERATORS

Probably the most important application of the behaviour of charged particles in electric and magnetic fields lies in the construction of accelerators to produce beams of high-energy particles which are used for many types of scattering experiments. The types of accelerators may be classified as follows:

(*i*) Direct or potential drop accelerators in which the particles are accelerated along a tube due to the action of a constant electric field. The electric field may be produced by an electrostatic generator or a voltage multiplier.

(*ii*) Orbital accelerators in which a combination of electric and magnetic fields is used to accelerate particles moving in a circular or spiral path.

(*iii*) Linear accelerators in which the particles move in a straight line path and are accelerated by an oscillating electric field. In one type of linear accelerator, the electric field generates an electromagnetic wave which carries the particles along, and in another form the electric field accelerates the particles as they pass across gaps at intervals along the accelerator tube.

Some accelerators of the orbital type are described below.

(*a*) *The cyclotron.* The basic features of the cyclotron are illustrated in Fig. 8.1. The source of positive ions is located at the mid-point of the gap between the two flat semicircular metal boxes which are called *dees* because of their D-shape. These dees are in fact electrodes with a high-frequency

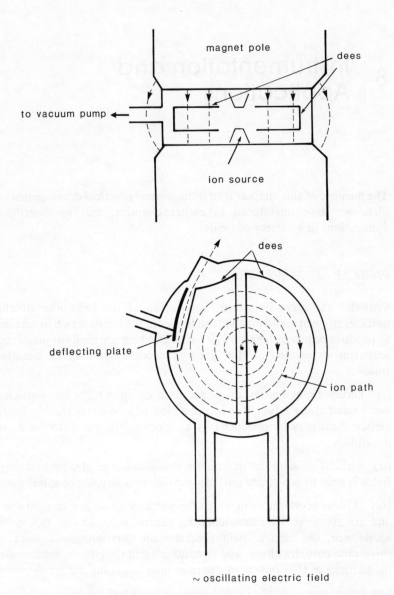

*Figure 8.1.* The basic features of the cyclotron.

oscillating electric field $E$ applied to them. This arrangement has the property that the electric field acts only across the gap and not inside the dees. When an ion of charge $+q$ is emitted from the ion source it is accelerated across the gap into one of the dees where the effect of the magnetic field constrains it to move in a circle. The equation of motion of the ion in the dee is (see Chapter 3) given by

$$\frac{mv^2}{r} = Bqv, \tag{8.1}$$

i.e.

$$r = \frac{mv}{Bq}, \tag{8.2}$$

and

$$\omega = \frac{v}{r} = \frac{Bq}{m}. \tag{8.3}$$

If, when the ion has completed a semicircle and reached the gap again, the electric field has changed direction, the ion will be accelerated across the gap once more so that the kinetic energy and the momentum $mv$ of the ion are increased. Then, from eqn (8.2), the radius of the orbit of the ion in its next semicircle is increased, and hence the ion follows a spiral path, receiving an acceleration each time it crosses the gap until it reaches the maximum possible radius $R$ when it is deflected out of the dee. However, the ion will only be accelerated each time it crosses the gap if the angular frequency $\omega$ of the ion's motion matches the angular frequency $2\pi f$ of the electric field. Using eqn (8.3) this condition can be written as

$$f = Bq/2\pi m, \tag{8.4}$$

and the frequency $f$ and the magnetic field $B$ must be chosen to satisfy this for a particular value of $q/m$.

The kinetic energy with which the ions leave the accelerator can be obtained from eqn (8.1) by putting the radius $r$ equal to the maximum radius $R$. Using non-relativistic kinematics this gives

$$\frac{1}{2}mv_{max}^2 = \frac{R^2B^2q^2}{2m} \tag{8.5a}$$

$$= 2\pi^2f^2mR^2. \tag{8.5b}$$

From these equations it follows that at non-relativistic energies the limitation on the maximum energy obtainable for a particular ion is determined by the properties of the cyclotron, in particular the size of the dees and the pole faces of the magnet. It is not possible, however, to increase the energy indefinitely by increasing the radius $R$ because the relativistic increase in mass begins to have an effect on the motion of the ions. Eqns (8.1)–(8.4) have general validity if we take $m$ to be the moving mass. Equation (8.3) can be rewritten in terms of the rest mass $m_0$ using eqn (2.7) as

$$\omega = \frac{Bq}{m_0}\left(1 - \frac{v^2}{c^2}\right)^{\frac{1}{2}},$$

which shows that as the ions are accelerated the mass increases and the angular frequency decreases. Thus, the ions traverse each semicircle too slowly and become out of phase with the electric field until eventually they cease to be accelerated. In order to attain relativistic energies it is necessary to vary either the frequency $f$ or the magnetic field $B$ in order to satisfy eqn (8.4). This is not done in the cyclotron, but is done in the synchrocyclotron and the synchrotron.

(b) *The synchrocyclotron.*   The synchrocyclotron resembles the cyclotron in every respect, except that the frequency $f$ of the electric field is allowed to change with time so that it remains in step with the angular frequency $\omega$. The ions still follow a spiral path but bunch together in groups. At very high energies the cost of manufacturing a sufficiently large magnet is prohibitive.

(c) *The synchrotron.*   In this machine the ions move in circular orbits of fixed radius. This has the great advantage that the magnetic field is required only in the region of the orbit and not over the whole area. In order to calculate the angular frequency and the magnetic field required in this case we must use the relativistic expression (2.11) for the total energy,

$$E^2 = p^2c^2 + m_0^2c^4, \tag{8.6}$$

so that the momentum is given by

$$p = mv = \frac{E}{c}\left\{1 - \left(\frac{m_0c^2}{E}\right)^2\right\}^{\frac{1}{2}}.$$

Now, rearranging eqn (8.2), the magnetic field required at a fixed radius $r_0$ is given by

$$B = \frac{E}{cqr_0}\left\{1 - \left(\frac{m_0c^2}{E}\right)^2\right\}^{\frac{1}{2}}, \tag{8.7}$$

and from eqn (8.3) the angular frequency of the ions at this radius is

$$\omega = \frac{c}{r_0}\left\{1 - \left(\frac{m_0c^2}{E}\right)^2\right\}^{\frac{1}{2}}. \tag{8.8}$$

*Table 8.1.*   Some operating characteristics of the electron synchrotron in the SERC
Daresbury Laboratory

| | |
|---|---|
| *Injector* (electron linear accelerator) | |
| Maximum energy | 10–15 MeV |
| Electron current | 20 mA |
| Frequency of r.f. field | 3 GHz |
| | |
| *Booster synchrotron* | |
| Maximum energy | 600 MeV |
| Electron current | 20 mA |
| Frequency of r.f. field | 500 MHz |
| Maximum magnetic field | 0.786 T |
| Number of electron bunches | 53 |
| | |
| *Storage ring* | |
| Maximum energy | 2 GeV |
| Electron current | 200–300 mA |
| Frequency of r.f. field | 500 MHz |
| Maximum magnetic field | 1.2 T |
| Number of electron bunches | 160 |
| Bunch spacing | 2 ns |
| Synchrotron radiated power | 51 kW at 200 mA |

r.f. = radio frequency.

The ions are again accelerated by the electric field as they cross gaps at
intervals along the circular path, and the frequency of this field is varied with
time to keep in step with the angular frequency $\omega$. The magnetic field also
varies with time to keep the particles in a fixed orbit. It is advantageous to
inject the charged particles at a relatively high energy, of the order of MeV,
so that the injection system itself comprises a small accelerator.

Synchrotrons can be used to accelerate either protons or electrons. Some
information about the 2 GeV electron synchrotron at the SERC Daresbury
Laboratory in the UK is given in Table 8.1 and Fig. 8.2.

## SYNCHROTRON RADIATION

In Chapter 2, we mentioned the electromagnetic radiation which is emitted
by an accelerated charged particle and is known as synchrotron radiation.
For a particle of rest mass $m$ moving in a circular path with energy $E$, the
energy loss per turn is proportional to $(E/mc^2)^4$, from which it is clear that
synchrotron radiation by electrons is much more important than for

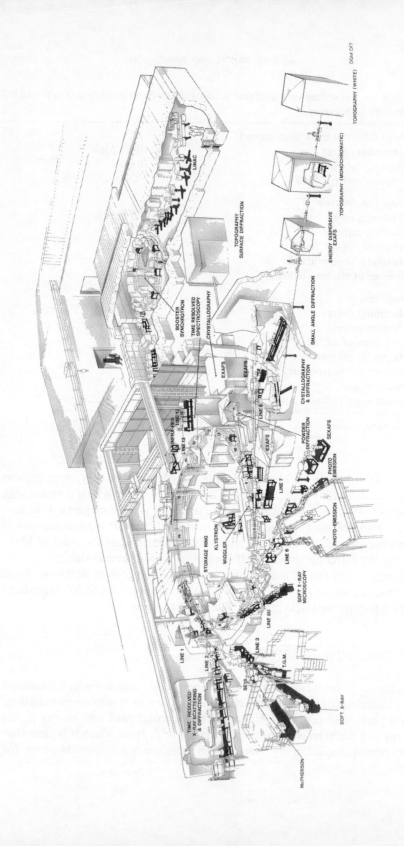

TOPOGRAPHY (WHITE)

TOPOGRAPHY (MONOCHROMATIC)

ENERGY DISPERSIVE EXAFS

SMALL ANGLE DIFFRACTION

CRYSTALLOGRAPHY & DIFFRACTION

POWDER DIFFRACTION

PHOTO SEXAFS

PHOTO-EMISSION

SOFT X-RAY MICROSCOPY

SOFT X-RAY

McPHERSON

T.G.M.

SEYA

LINE 5U

LINE 3

LINE 2

LINE 1

TIME RESOLVED X-RAY SCATTERING & DIFFRACTION

STORAGE RING

WIGGLER

KLYSTRON

LINE 6

LINE 7

EXAFS

LINE 8

EXAFS

LINE 9

EXAFS

CRYSTALLOGRAPHY

INFRA-RED LINE 12

LINE 13

TIME RESOLVED SPECTROSCOPY

BOOSTER SYNCHROTRON

TOPOGRAPHY SURFACE DIFFRACTION

LINAC

DGM 017

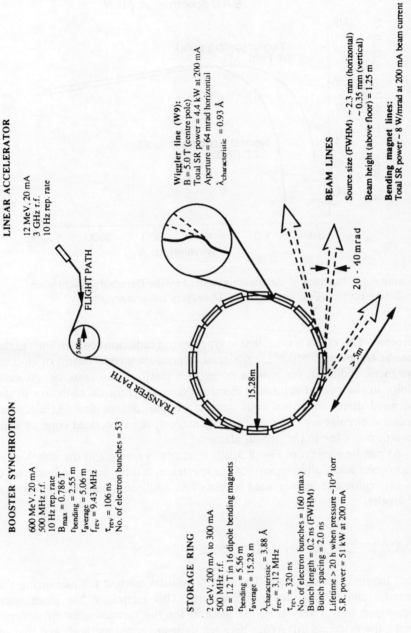

**BOOSTER SYNCHROTRON**

600 MeV, 20 mA
500 MHz r.f.
10 Hz rep. rate
$B_{max}$ = 0.786 T
$r_{bending}$ = 2.55 m
$r_{average}$ = 5.06 m
$f_{rev}$ = 9.43 MHz

$\tau_{rev}$ = 106 ns
No. of electron bunches = 53

**LINEAR ACCELERATOR**

12 MeV, 20 mA
3 GHz r.f.
10 Hz rep. rate

FLIGHT PATH

5.06 m

TRANSFER PATH

15.28 m

20 - 40 mrad

> 5 m

**STORAGE RING**

2 GeV, 200 mA to 300 mA
500 MHz r.f.
B = 1.2 T in 16 dipole bending magnets
$r_{bending}$ = 5.56 m
$r_{average}$ = 15.28 m
$\lambda_{characteristic}$ = 3.88 Å
$f_{rev}$ = 3.12 MHz

$\tau_{rev}$ = 320 ns
No. of electron bunches = 160 (max)
Bunch length = 0.2 ns (FWHM)
Bunch spacing = 2.0 ns
Lifetime > 20 h when pressure ~$10^{-9}$ torr
S.R. power = 51 kW at 200 mA

**Wiggler line (W9):**
B = 5.0 T (centre pole)
Total SR power = 4.4 kW at 200 mA
Aperture = 64 mrad horizontal
$\lambda_{characteristic}$ = 0.93 Å

**BEAM LINES**

Source size (FWHM) ~ 2.3 mm (horizontal)
~ 0.35 mm (vertical)
Beam height (above floor) = 1.25 m

**Bending magnet lines:**
Total SR power ~ 8 W/mrad at 200 mA beam current

*Figure 8.2.* The layout of the 2 GeV synchrotron radiation source at the SERC Daresbury Laboratory. (Courtesy of the SERC Daresbury Laboratory.)

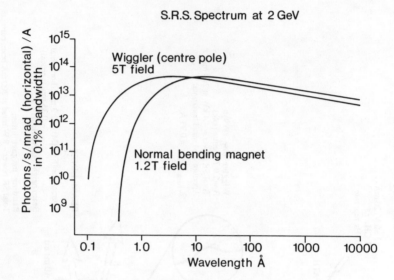

*Figure 8.3.* Intensity of the photons emitted by the Daresbury synchrotron radiation source. (Courtesy of SERC Daresbury Laboratory.)

protons. This effect is exploited in synchrotron radiation sources, such as the one at Daresbury, which are designed to produce intense beams of electromagnetic radiation for a wide range of scientific investigations. At each point in the curved orbit, the electrons emit synchrotron radiation in the forward direction into a cone centred on the tangent line. As electrons traverse circular arcs in each bending magnet, the associated cone of light sweeps out a fan in the orbital plane.

As can be seen from Fig. 8.3, the intensity is greatest in the soft X-ray, ultraviolet and visible regions of the spectrum, but the intensity in the hard X-ray region can be increased by use of an additional magnet, known as a "wiggler".

## MASS SPECTROMETRY AND MASS SEPARATION

In Chapter 3 we described Thomson's original method for measuring the specific charge $q/m$ for positive ions. This technique has since been developed into a very accurate method for the measurement of isotopic mass; the principal difference between these advanced methods and that of Thomson is that all ions with the same $q/m$ are brough to focus on a line or

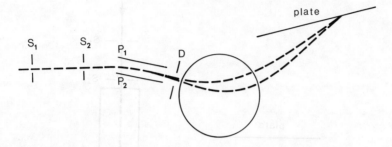

*Figure 8.4.*  Diagram of the first mass spectrograph designed by Aston.

point, instead of being spread out in a parabola, and there is greater dispersion or separation of ions of different $q/m$.

A diagram of the *mass spectrograph* designed by Aston is shown in Fig. 8.4. The positive ions pass through the slit system $S_1S_2$ and are then deflected by the electric field between the plates $P_1$ and $P_2$, which gives a deflection proportional to $1/v^2$. Thus, a beam which initially has a spread of velocities is broadened as it passes through the electric field. A group of these ions is selected by a wide diaphragm D and passes through a magnetic field which again deflects the slower ions more than the faster ions. At some point outside the field the paths of the fast and slow ions with the same $q/m$ intersect, and so these ions can be brought to a focus on the photographic plate. Ions with the same range of energies but different $q/m$ come to a focus at a different point. This method of producing a focus is called *velocity focusing*.

In the mass spectrograph the mass spectrum appears as a series of lines on a photographic plate. In a *mass spectrometer* the ions are focused on a slit, which is more convenient for measuring abundances.

A mass spectrometer designed by Bainbridge following the method of Dempster is shown in Fig. 8.5. Singly charged positive ions produced in the space above the slit $S_1$ are accelerated towards $S_2$ by an electric field, and emerge with a range of velocities. The ions then pass through crossed electric and magnetic fields which select those ions with velocity $v = E/B_1$ and allow them to pass through slit $S_3$. Beyond $S_3$ the ions pass into a region where a magnetic field $B_2$ acts perpendicular to the plane of the figure, so that the ions traverse a semicircle with radius $R = mv/qB_2$. Thus, the radius depends linearly on the mass, so that ions of different mass strike the plate at different points and yield a record of the spectrum of masses present in the beam. This method of focusing the ions is known as *direction focusing*. The photographic place can be replaced by an ion collector which records the intensity of the current due to the ions, and comparison of the relative ion currents

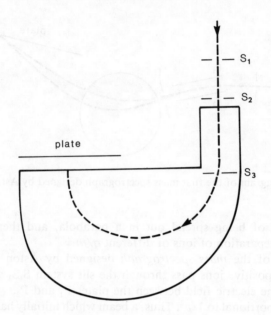

*Figure 8.5.* A version of the mass spectrometer designed by Bainbridge.

gives a measure of the relative abundance of the various isotopes.

Further development of the mass spectrometer may be of a *single-focusing* type and usually involves a magnetic analyser, which is a wedge-shaped magnet with an angle of about 60°. Instruments which achieve both focusing of ions with different initial velocities and different initial directions are known as *double-focusing* devices. They usually involve a magnetic analyser and an electrostatic analyser of cylindrical form.

The ISOLDE mass separators now in use at the European Nuclear Research Centre, CERN, provide isotope separation on line. They are attached to the 600 MeV proton synchroton and the impact of the high-energy protons on selected targets yields a variety of nuclear reaction products which emerge as singly charged ion beams at 60 keV. In ISOLDE-2, shown on the left of Fig. 8.6, the ion beam is directed on to an electrostatic lens system consisting of four cylindrical electrodes which a produce a parallel beam directed into a 55° focusing magnetic analyser. The individual ion beams can be selected and distributed to experimental areas, as shown in Fig. 8.6. Some 67 elements can be separated and can be used for an extensive programme of research in nuclear, atomic, solid-state and surface physics.

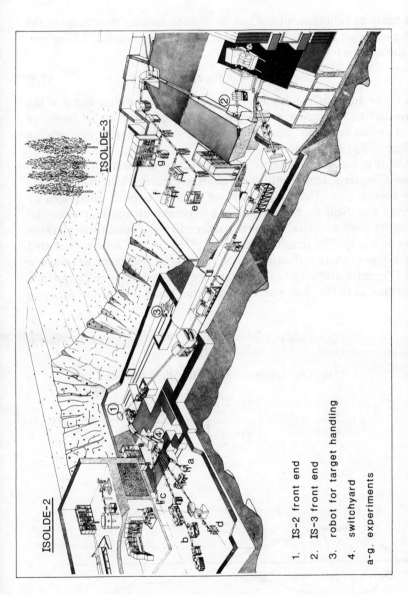

*Figure 8.6.* The on-line isotope mass separators ISOLDE. The synchrocyclotron (beyond the right of the picture) delivers beams of 600 MeV protons or 900 MeV $^3$He ions to ISOLDE-2 (1974) or to ISOLDE-3 (1987). The two separators start at the points marked 1 and 2, respectively, on the diagram. At point 3 is shown the robot for target handling and at point 4 the switchyard which distributes the separated beam to various experiments, which are located at the points marked a–d. Other experiments are located at e–g. (Courtesy of CERN, Geneva.)

1. IS-2 front end
2. IS-3 front end
3. robot for target handling
4. switchyard
a-g. experiments

ISOLDE-2

ISOLDE-3

## ELECTRON LENSES

When a beam of light passes from one medium to another the change in the refractive index causes the beam to change its original direction. The refraction at the surface is determined by Snell's Law,

$$\sin i = \mu \sin r, \tag{8.9}$$

where $i$ is the angle of incidence, $r$ is the angle of refraction and $\mu$ is the refractive index. This property is utilized in the construction of lenses of glass and similar materials; these lenses may have a converging or diverging effect depending on their design.

The ability of electric and magnetic fields to cause deviation of an electron beam (see Chapter 3) can be utilized to construct lenses which can focus the electron beams. The effect of an electrostatic field is most easily seen by considering the field to be composed of regions of constant potential separated by lines of equal potential, rather like the system of marking contours on a map. The smooth continuous effect of the field is represented by a succession of sharp refractions at the equipotential surfaces, as shown in Fig. 8.7. The refractive effect can be examined by considering refraction at a single surface, as in Fig. 8.8, where the velocities in the two regions are $v$, $v'$, and

$$\sin i = \frac{v_x}{v}, \quad \sin r = \frac{v'_x}{v'}, \tag{8.10}$$

but $v_x = v'_x$ since there is no acceleration in the $x$-direction. Hence

$$\mu = \frac{\sin i}{\sin r} = \frac{v'}{v} = \left( \frac{2V_2 \frac{e}{m}}{2V_1 \frac{e}{m}} \right)^{\frac{1}{2}} = \left( \frac{V_2}{V_1} \right)^{\frac{1}{2}}. \tag{8.11}$$

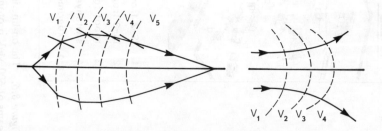

*Figure 8.7*  A representation of the converging or diverging effect of an electrostatic field by successive refractions at equipotential surfaces.

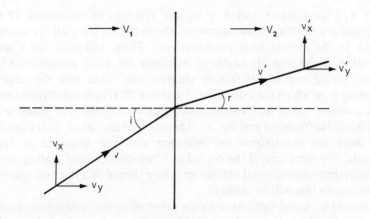

*Figure 8.8.*   Refraction at a single surface.

Thus, if electrons pass into regions of higher potential there is a focusing effect, and if they pass into a region of lower potential there is a diverging effect, as shown in the two parts of Fig. 8.7.

## X-RAY AND ELECTRON DIFFRACTION

We now examine some applications of wave behaviour. One of the most important of these applications is the study of crystalline structures by means of diffraction of X-rays and electrons. We have already noted that the de Broglie wavelength of electrons lies in the X-ray region, and that such wavelengths are comparable with the spacing of atoms or ions in crystals.

When radiation falls on an atom (or ion) the orbital electrons are set into forced vibration and the atom emits radiation of the same frequency. This secondary radiation is emitted in all direction. In Fig. 8.9 it is assumed that a plane wave is incident at an angle $\theta$ to the crystal surface, and the scattered radiation is examined at a particular angle $\theta'$. The condition for constructive interference between radiation scattered from atoms A and B is that the path difference $A'B - AB'$ is equal to an integral number of wavelengths. Similarly, the condition for constructive interference between radiation scattered from atoms B and C is that the path difference $B''C + CC'$ is equal to an integral number of wavelengths. These two conditions lead to the *Laue equations*,

$$b (\cos \theta - \cos \theta') = k\lambda, \tag{8.12a}$$

$$a (\sin \theta + \sin \theta') = l\lambda, \tag{8.12b}$$

where $k$, $l$ are integers and $a$, $b$ are the spacings of the atoms. If these equations are satisfied simultaneously, the radiation from all the scattering centres in the crystal adds constructively. Thus, solutions for $\theta$ and $\theta'$ determine all the possible angles of incidence for which constructive interference giving rise to diffraction maxima can occur and the angles of scattering $\theta'$ at which they do occur. Equation (8.12a) is exactly the same as the condition for a diffraction maximum to be observed using a one-dimensional diffraction grating, and the essential feature of such a grating is that there are transparent or reflecting elements arranged at regular intervals. We have treated the crystal as a two-dimensional grating; it is, of course, a three-dimensional grating and there should be a third equation, but for simplicity this will be omitted.

It should be noted that, for a single plane of atoms, scattering can occur for cases in which the angle of incidence $\theta$ is not equal to the angle of reflection $\theta'$. It can be shown, however, that any such solution of eqns (8.12a) and (8.12b) can be interpreted in terms of specular reflection, i.e. angle of incidence equal to angle of reflection, from some other set of planes through the crystal lattice. Consider the plane MN drawn through the atom C in Fig. 8.9 in such a way that the angle of incidence at this plane is equal to the angle of reflection. If this angle is denoted by $\phi$ we have

$$\phi = \theta + \alpha = \theta' - \alpha,$$

or

$$\theta = \phi - \alpha, \quad \theta' = \phi + \alpha, \quad \alpha = \tfrac{1}{2}(\theta - \theta'),$$

where $\alpha$ is the angle between the plane MN and the crystal surface. Substituting for $\theta$, $\theta'$, eqns (8.12a) and (8.12b) become

$$2b \sin \phi \sin \alpha = k\lambda \tag{8.13a}$$

$$2a \sin \phi \cos \alpha = l\lambda. \tag{8.13b}$$

Now, we let the integer $n$ be the highest common factor of $k$ and $l$, so that

$$k = nk_1, \quad l = nl_1,$$

and eqns (8.13a) and (8.13b) can be written as

$$\sin \alpha = \frac{n\lambda}{2 \sin \phi} \frac{k_1}{b}, \quad \cos \alpha = \frac{n\lambda}{2 \sin \phi} \frac{l_1}{a}. \tag{8.14}$$

Then, squaring and adding, we have

$$1 = \left(\frac{n\lambda}{2 \sin \phi}\right)^2 \left(\frac{k_1^2}{b^2} + \frac{l_1^2}{a^2}\right),$$

or

$$2d \sin \phi = n\lambda, \tag{8.15}$$

where $d$ has the dimensions of length and is defined by

$$d = \left( \frac{k_1^2}{b^2} + \frac{l_1^2}{a^2} \right)^{-\frac{1}{2}}. \tag{8.16}$$

Equation (8.15) shows that the phenomenon of diffraction by atoms in a crystal can be interpreted in terms of specular reflection from the plane MN. This representation is due to W. L. Bragg (1913) and the phenomenon is known as *Bragg reflection*. The integer $n$ is known as the order of reflection. It must be noted, however, that what occurs is not specular reflection of all wavelengths but reflection of selected wavelengths which satisfy eqn (8.15). Further, the reflection is a spatial and not a surface effect, since reflection occurs at all planes parallel to MN and characterized by the indices $k_1$ and $l_1$ and, because the reflected intensity for a single plane is small, it is the constructive contribution from all these planes which leads to an observable intensity for the diffraction maxima.

The plane MN is known as a lattice plane or net plane for the crystal. Using eqn (8.14), the angle of the plane to the surface of the crystal is given by $\tan \alpha = k_1 a / l_1 b$, so that the spacing of atoms in the plane is determined by the indices $k_1$ and $l_1$. In the special case of $k_1 = 0$, $l_1 = 1$, we have

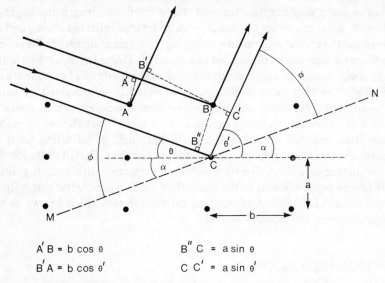

| | |
|---|---|
| $A'B = b \cos \theta$ | $B''C = a \sin \theta$ |
| $B'A = b \cos \theta'$ | $CC' = a \sin \theta'$ |

*Figure 8.9.*   The diffraction of X-rays from a crystal, which appears as specular reflection from the plane MN.

$\alpha = 0$ and $\theta = \theta'$, so that the reflection planes are parallel to the surface, and from eqn (8.16) we have $d = a$. Hence, for this special case we may interpret $d$ as the spacing between the lattice planes. This interpretation is generally true but the general proof will not be given here.

The experimental study of crystal structures involves the measurement of the spacing between the planes and the interpretation of the results in terms of the spacing and arrangement of the atoms in the crystal. (The arrangement in real crystals is much more complicated than the simple cubic array we have used in Fig. 8.9.) The patterns observed in diffraction experiments depend on the experimental procedure. If, for example, a wide spectrum of X-ray radiation is used, each wavelength can be reflected at a different set of lattice planes at an angle of reflection determined by the Laue equations, and for a fixed angle of incidence a complicated pattern known as the *Laue spot pattern* is observed on a photographic plate.

Figure 8.10a shows the X-ray diffraction pattern from a fibre of DNA (deoxyribonucleic acid). Fibres consist of an array of microcrystals randomly oriented about one axis so that the Bragg reflection condition is always satisfied and a single X-ray wavelength can be used. Imperfections in the arrangement of the microcrystals lead to elongation of the diffraction spots into arcs. Figure 8.10b shows part of the first fibre diffraction picture to be recorded from DNA using neutrons. The similarity of Figs 8.10a and b gives yet further confirmation of the wave–particle duality.

A simpler and more easily interpreted result is obtained if monochromatic radiation and a single crystal are used. For a fixed wavelength and angle of incidence, a particular set of planes gives an intense reflected beam, and so by rotating the crystal and thereby changing the angle of incidence the reflection from various sets of planes can be examined. This method, known as the Bragg spectrometer method, yields the most accurate results but is laborious and slow. The ring patterns illustrated in Fig. 3.9 are obtained by using monochromatic radiation and a polycrystalline specimen, which contains many small crystals randomly oriented. This means that the set of planes which yield reflected radiation at a particular angle of reflection lie at all possible orientations to the incident beam, and so the reflected radiation lies on the surface of a cone. The intersection of each cone with a photographic plate placed perpendicular to the beam then yields a concentric ring pattern on the plate. This method of studying diffraction patterns is known as the *Debye–Scherrer* and *powder method*.

## ELECTRON MICROSCOPE

In the optical microscope a system of glass lenses is used to produce a magnified image which may be viewed by eye or projected on to a screen or

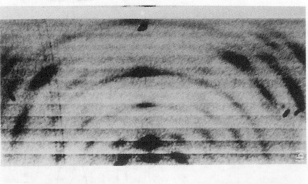

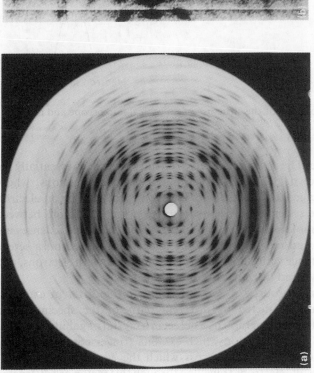

*Figure 8.10.* (a) X-ray diffraction pattern from a fibre of DNA (deoxyribonucleic acid) recorded on photographic film at the SERC Daresbury Laboratory Synchrotron Radiation Source. The X-ray wavelength was 1.5 Å. (*Courtesy of Dr A. Mahendrasingam, Dr V. T. Forsyth and Professor W. Fuller, Department of Physics, University of Keele.*) (b) Part of a neutron diffraction pattern recorded from a fibre of DNA (deoxyribonucleic acid) at the Institut Laue-Langevin, Grenoble, high flux reactor neutron source. The neutron wavelength was about 2.4 Å. (*Courtesy of Dr V. T. Forsyth, Dr A. Mahendrasingam and Professor W. Fuller, Department of Physics, University of Keele.*)

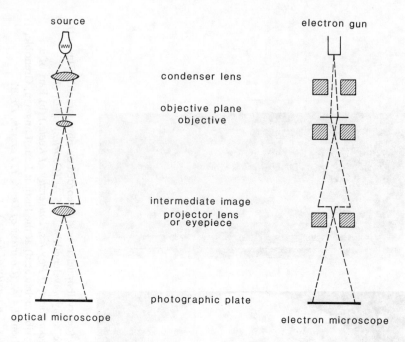

*Figure 8.11.* Comparison of the structure of an optical microscope and an electron microscope.

photographic plate. The *transmission* electron microscope has essentially the same form, as shown in Fig. 8.11. The initial beam of electrons, whose energy is in the range from 100 keV to 3 MeV, is produced by an electron gun, and the whole system must be in vacuum to prevent collisions between the electrons and molecules in the air. Electrostatic or magnetic lenses are used for focusing and magnification. The image is produced on a screen coated with a substance with fluoresces when bombarded with electrons, and the screen can be viewed directly or photographed.

The most important source of contrast is due to the different extent to which electrons are diffracted by different parts of the specimen. An example of an electron micrograph of a defect in a crystal of gallium arsenide is shown in Fig. 8.12. The region which exhibits fringes corresponds to a portion of a plane in the crystal across which the atoms are "stacked" incorrectly. As a result, electrons travelling through this region are diffracted differently and give rise to interference fringes.

For optical instruments the resolving power (i.e. the ability to resolve or distinguish between two points close together) is given by $0.61\lambda/\sin\alpha$ where $\lambda$ is the wavelength in the medium in which the object is situated and $\alpha$ is the angular aperture of the objective lens. The same formula applies to the

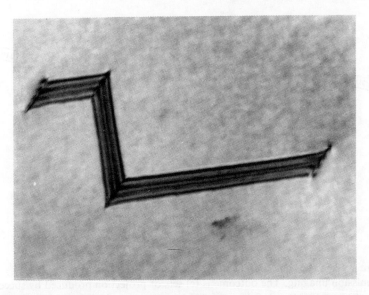

*Figure 8.12.*   A transmission electron microscope image of a defect in a crystal of gallium arsenide (GaAs). This defect is called a stacking fault and consists of a plane sloping from top to bottom of the crystal across which the atoms are incorrectly stacked. The electrons passing through the crystal intersect the stacking fault at different depths and give rise to a fringe contrast. The magnification is about 40 000 × . (Courtesy of Dr P. Charsley and Dr U. Bangert, Department of Physics, University of Surrey.)

electron microscope where $\lambda$ is the wavelength given by the de Broglie relation, $\lambda = h/p$. The lens apertures which can be used in electron microscopes are restricted by practical limitations, but the small value of the electron wavelength means that the resolving power of the electron microscope is several thousand times greater than the optical microscope.

Efforts are being made to increase the electron energies into the MeV region and so to penetrate thicker specimens.

In the *scanning* electron microscope only a point on the axis of the lens system is illuminated. This means that the lens system need provide good focusing only on the axis. This has made possible the extension of scanning microscopy to acoustic and X-ray beams.

## X-RAY RADIOGRAPHY AND TOMOGRAPHY

When a beam of X-rays is passed through a complex object the differences in the attenuation coefficients of the different elements in the object cause

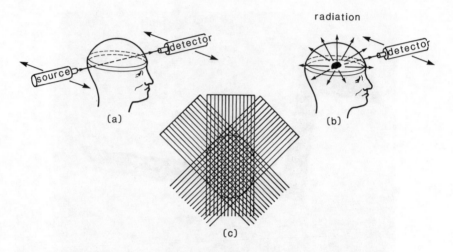

*Figure 8.13.* The tomographic procedure for (a) transmission imaging or
(b) emission imaging. The outcome is (c) sets of projection produced by successive
translation and rotation of the source/detector system. (Courtesy of The Institute
of Physics. From R. A. Brooks and G. diChiro, *Phys. Med. Biol.* **21**, 689, 1976.)

different amounts of absorption (see Chapter 2 and Fig. 2.10). If a film is
placed behind the object, the different attenuation leads to different
amounts of blackening, and hence the picture, called a *radiograph*, dif-
ferentiates between different materials. Thus, it is possible to see the outline
of a crack in a metal plate or of a broken bone in tissue.

For medical purposes it is particularly valuable to reconstruct the variation
of attenuation in a layer or body section and to do this for many successive
layers in order to build up a three-dimensional picture. This technique is
called *tomography* and the essential features are shown in Fig. 8.13a. The
X-rays traverse only the layer under examination. A set of views or
*projections* can be obtained by first translating and then rotating the detector
and source, and these give sufficient information to allow the determination
of the X-ray attenuation coefficient in each volume element of the layer. The
reconstruction of the image is usually done automatically by a small
computer.

Figure 8.14a shows an X-ray tomograph of a slice through a tree-trunk.
The tree rings show up clearly because of the variation in attenuation
coefficient shown in Fig. 8.14b.

*Figure 8.14.* (a) A X-ray tomograph of a slice through a tree-trunk showing a
crack and a stress pattern due to branch formation. (b) The variation of
attenuation coefficient of a diameter shows why the tree rings show up so clearly.
(Courtesy of The Institute of Physics. From D. F. Jackson, J. Foster, W. B.
Gilboy, K. Kouris and N.M. Spyrou, *Inst. Phys. Conf. Ser.* **44**, 95, 1979.)

(a)

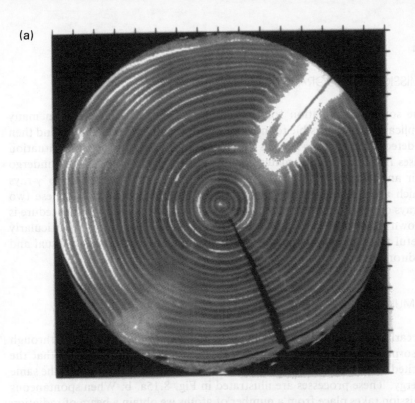

−453    −173

(b)

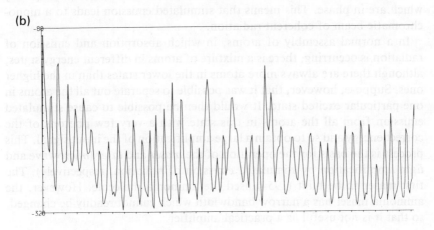

TREE TEST

21.12.77

## EMISSION TOMOGRAPHY

The source of radiation for tomography need not be external; in many applications it is possible to inject or insert a radioactive substance and then to detect the emitted $\gamma$-rays, as shown in Fig. 8.13b. An interesting situation arises if the radioactive material emits positrons. These positrons undergo pair annihilation with an electron in the material to produce two $\gamma$-rays which are emitted almost "back-to-back" (see Problem 2.8). These two $\gamma$-rays can be detected in coincidence, i.e. simultaneously. This procedure is known as *positron emission tomography*, which has proved particularly useful for studying the activity of the brain and its response to visual and auditory stimuli.

## STIMULATED EMISSION OF RADIATION

In earlier chapters we have examined the excitation of an atom through absorption of a photon of the appropriate energy and we know that the excited atom may then radiate spontaneously and emit a photon of the same energy. These processes are illustrated in Fig. 8.15a, b. When spontaneous emission takes place from a number of atoms we obtain a beam of radiation of observable intensity but there is no particular phase relation between the wave motions associated with the photons emitted by different atoms, and the beam is said to be *incoherent*. It is also possible that a photon of the correct energy can stimulate an excited atom to emit another photon and make a transition to its lower state. This process of stimulated emission is illustrated in Fig. 8.15c, and leads to two photons of the same frequency which are in phase. This means that stimulated emission leads to a monochromatic beam of coherent radiation.

In a normal assembly of atoms, in which absorption and emission of radiation is occurring, there is a mixture of atoms in different energy states, although there are always more atoms in the lower states than in the higher ones. Suppose, however, that it was possible to separate out all the atoms in one particular excited state. It would then be possible to cause stimulated emission from all the atoms in this state with a very few photons of the correct energy and so to obtain a large amplification of the input signal. This process is the basis for the operation of the *maser* and *laser* (*m*icrowave and *l*ight *a*mplification by *s*timulated *e*mission of *r*adiation, respectively). The first maser to be built (1954) used the ammonia molecule. However, the ammonia maser has a narrow bandwidth which cannot readily be changed, so that it is not useful as a practical amplifier.

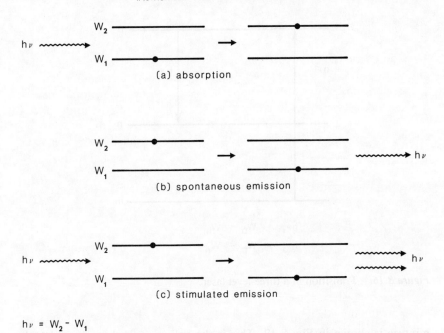

$$h\nu = W_2 - W_1$$

*Figure 8.15.* Comparison of the transitions leading to (a) absorption of a photon, (b) spontaneous emission of a photon, and (c) stimulated emission of a photon.

The addition of energy from outside is known as *pumping*. An example of this process involving three energy levels is illustrated in Fig. 8.16. Atoms are excited from the ground state m to the highest state p by radiation of energy $h\nu_1$, until the number of atoms in state p is much higher than in state n. A weak signal is introduced with frequency $\nu_2$ such that the energy $h\nu_2$ is equal to the energy gap between states p and n, and will cause stimulated emission so that the input signal is amplified. Spontaneous emission from state n to state m will repopulate the lower state so that the process of pumping and amplification can proceed simultaneously and the laser can operate continuously.

## THE HELIUM-NEON LASER

Laser operation has been obtained with many different types of solid-state materials and with mixtures of gases. An example of a laser using a gas

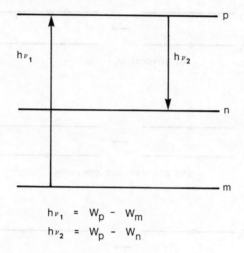

$$h\nu_1 = W_p - W_m$$
$$h\nu_2 = W_p - W_n$$

*Figure 8.16.*   Transition in a three-level laser.

mixture is shown in Fig. 8.17. This is the helium–neon laser in which the helium atoms are excited by collisions with electrons in a gas-discharge tube. Energy is transferred to neon atoms by collisions (see Chapter 5 and Fig. 5.4b). Laser emission is associated with the initial de-excitation of the neon atoms. The neon atoms then make further spontaneous de-excitations to their ground state.

An important feature of this system is the pair of reflecting plates at each end, as can be seen in Fig. 8.17; the output signal builds up in intensity as the emitted photons are reflected between these plates until the amplification is great enough for the beam to pass through one of the partially silvered end plates. Photons which are emitted at an angle to the axis of the laser are lost through the sides of the system, so that the final beam is a coherent wave with plane wavefronts perpendicular to the axis of the laser.

The output from lasers has been used for many interferometer experiments, including a check on the Michelson–Morley experiment. The intense local heating produced by the beam can be used for many unusual types of heating and welding applications, such as restoration of a detached retina in a human eye. As an amplifier, the low noise produced makes the maser very useful in situations where weak signals are being detected, and masers have been used to extend the range and resolution of radio telescopes. Finally, the laser is useful in the field of long-distance communications in conjunction with optical fibres.

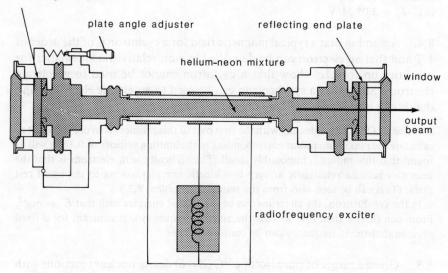

*Figure 8.17.*    The helium–neon gas laser.

## PROBLEMS

**8.1.**    A cyclotron whose magnetic pole faces have a diameter of 60 inches is operated with a radio-frequency field at $10^7$ Hz. Show that this machine can accelerate protons to an energy of approximately 10 MeV and $\alpha$-particles to approximately 40 MeV. (It is customary to give the dimensions of cyclotrons in inches. The required conversion to SI units is 60 inches = 1.52 m.)

**\*8.2.**    It is required to accelerate a beam of singly charged lithium ions to an energy of 20 MeV. Determine a suitable set of cyclotron parameters, i.e. pole face diameter, rf frequency, magnetic field strength. Make a list of the ancillary equipment necessary to operate this machine and carry out experiments, and estimate the minimum size of the building needed to house it.

**8.3.**    Calculate the kinetic energy when the velocity is $0.98c$ of:

(*i*)    a proton;

(*ii*)    an electron.

*Solution.* (i)  $T_p$ = 3.76 GeV
(ii)  $T_e$ = 2.05 MeV

**8.4.** Assuming that a typical magnetic field for a cyclotron is of the order of 1 T and that an electron can be described by non-relativistic kinematics for velocities up to 0.1c, show that a cyclotron cannot be used to accelerate electrons. Show that a synchrotron can be used to accelerate electrons, and that it can be operated with a constant rf frequency.

*Solution.* One way of dealing with the first part of this problem is to use eqn (8.2) to calculate the maximum radius corresponding to the limiting velocity of 0.1c. It will be found that this radius is impossibly small. The difficulty with electrons is that the velocities become relativistic at very low kinetic energies, owing to the small rest mass. (This can be seen also from the results of Problem 8.3.)

In the synchrotron, the electrons can be injected at energies such that $E \gg m_0 c^2$. From eqn (8.8) it then follows that the angular frequency $\omega$ is constant for a fixed orbit so that the rf frequency can be constant in time.

**8.5.** Given a target of pure isotope $^{12}$C $(Z = 6)$ devise nuclear reactions with beams of protons or $\alpha$-particles which will lead to the production of isotopes of nitrogen $(Z = 7)$.

**\*8.6.** A beam of electrons with velocity $v$ enter a long solenoid at an angle $\theta$ to the axis of the solenoid. Show that the motion of the electrons is in the form of a helix of radius $mv \sin \theta / eB$ and pitch $2\pi mv \cos \theta / eB$, where $B$ is the magnetic field along the axis of the solenoid. Examine the behaviour of this system as a lens and show that its focal length is infinite.

*Solution.* The velocity of the electrons can be resolved into a component $v \cos \theta$ along the axis of the solenoid and parallel to $B$ and a component $v \sin \theta$ perpendicular to $B$. The presence of the latter component causes a circular motion around the lines of force, but at the same time the former component carries the electrons forward, so that the motion is like a screw-thread or helix. The radius of the helix is determined from the usual relation

$$\frac{m(v \sin \theta)^2}{r} = Be(v \sin \theta), \quad \text{i.e.} \quad r = \frac{mv \sin \theta}{eB}.$$

The angular frequency of rotation is

$$\omega = \frac{v \sin \theta}{r} = \frac{eB}{m},$$

so that the period of rotation is

$$T = \frac{2\pi}{\omega} = \frac{2\pi m}{eB}$$

and the distance traversed in the forward direction during one rotation is

$$s = v \cos \theta \, T = 2\pi m v \cos \theta / eB.$$

This is the pitch of the helix.

If a beam of electrons enter a range of angles $\theta$ such that $\cos \theta$ does not vary very much, electrons will cross the axis at the same point after a rotation, so that the solenoid has a limited focusing effect. However, if a beam of electrons enter parallel to the axis with $\theta = 0$ the field $B$ has no effect on them. Thus, a parallel beam remains parallel, and the focal length of the lens is infinite.

**8.7.**   Radiation is incident at an angle $\phi$ on two continuous reflecting planes separated by a distance $d$. Show that the condition for constructive interference of the reflected radiation is $2d \sin \phi = n\lambda$. Comment on the fundamental difference between this derivation of the Bragg reflection formula and that given in the text.

**\*8.8.**   A crystal has a basic unit consisting of atoms at the corners of a cube and another atom of the same type at the centre of the cube. Consider Bragg reflections from planes parallel to a cube face and show that diffraction maxima corresponding to odd orders of reflection are suppressed.

# Index

Accelerators, 183–187
Alpha-decay, 161, 167–169, *see also* Radioactivity
Alpha-particle, 68, 161
Ångström unit, 9
Angular frequency, 84, 121
Angular momentum, 80, 100
Antimatter, 35
Antiparticle, 34, 144
Atomic mass unit, 57
Atomic number, 108, 143
Atomic spectra
  absorption, 95
  combination principle, 78
  controlled excitation, 94
  effect of spin, 99
  in magnetic field, 103
  of alkali atoms, 96
  of helium, 106
  of hydrogen, 75–77
  selection rules, 78, 99
  stellar, 96, *see also* X-ray spectra
Atomic weights, 57
Atoms
  models of, 69–75, 79–81
  sizes of, 68
Attenuation, 7, 36
  linear coefficient, 36
  mass coefficient, 37

Balmer series, 76
Barn, 150
Bequerel, 165
Beta-decay, 161, 170–171, *see also* Radioactivity

Bindingt energy
  nuclear, 160–161, 176
  of electrons in atoms, 80
  per nucleon, 161
Bohr magneton, 101
Bohr's planetary model, 79–81
  deficiencies in, 109–110
Bohr's quantum postulates, 78
Bragg reflection, 197 *see also* X-ray diffraction
Brownian motion, 3

Cathode rays, 44
  specific charge of, 50
Centre of mass frame, 5
Chaos, 119
Classical mechanics, 26, 119
  deficiencies of, 119
Collisions, 3
  elastic, 3, 91, 150
  inelastic, 5, 91, 150
  ionization, 91 *see also* Conservation laws; Nuclear reactions; Scattering
Combination principle, 78
Complementarity, 62
Compton scattering
  comparison with photoelectric effect, 32
  conservation laws in, 29–30
  wavelength shift, 30
Compton wavelength, 31
Conservation
  of angular momentum, 171
  of energy, 28
  of linear momentum, 29

Conservation laws, 28
  for Compton scattering, 29
  for nuclear reactions, 150, 159
  for pair production, 34
Correspondence principle, 82
Critical potential, 92
Cross-section
  differential, 151
  for Rutherford scattering, 70, 72, 181
  total, 151
Curie, 165
Cyclotron, 183–186

Dalton's atomic hypothesis, 2
de Broglie wavelength, 58
Determinism, 119
Diffraction
  of electrons, 58, 195–198
  of neutrons, 199
  of X-rays, 195–198
Disintegration constant, 162
Dispersion, 13
Dispersive medium, 13, 76, 122
Distance of closest approach, 71
Dose, 166
Dose equivalent, 166

Electrolysis, 43
  laws of, 43
Electromagnetic radiation, 13–18
  classical theory, 13–15, 19
  dual nature, 39
  line width, 155
  velocity, 8, 16
Electromagnetic spectrum, 16–17
Electron
  charge, 8, 51
  fundamental constituent of matter, 51
  fundamental unit of charge, 54
  specific charge, 50
  spin, 99
Electron diffraction, 58, 195–198, see
    also X-ray diffraction
Electron gun, 45
Electron lens, 194
Electron microscope, 198–201
Electrons in metals, 34, 131
  energy bands, 134

energy gap, 134
Fermi energy, 132
quantized energies, 130
Electron volt, 9
Energy balance in nuclear reactions, 159
Energy level, 89
  width, 155
Energy level diagram, 89
  for carbon nucleus, 156
  for hydrogen atoms, 90
  for sodium atom, 91
Excitation potential, 92
Excited state, 89
Exclusion principle, 110

Faraday constant, 43
Faradays' laws, 43
Fermi energy, 132
Fission, 171–173
Frame of reference, 5
Frequency, 16, 120
Fundamental particles, 144
Fusion, 173

Gamma-ray, 17, 161
g-factor
  for atoms, 101
  for nucleus, 157
Gray, 166
Greek alphabet, 10
Ground state, 89
Group velocity, 122
  relation to particle velocity, 126
  relation to wave velocity, 128

Half-life, 162
Helium-neon laser, 205–207
Hybridization, 139
  tetragonal, 139
  trigonal, 140
Hydrogen spectrum, 75–77

Infrared radiation, 16, 20
Interference, 15
Ion, 43

Ionization energy, 89
   and electronic configuration, 113, 117
Ionization potential, 92
Isotope, 57
Iostopic weights, 57

Kinetic theory of gases, 3

Laboratory frame, 5
Lambert-Beer law, 7
Laser, 204
Laue equations, 195
Line spectrum, 75, *see also* Atomic
   spectra
Liquid drop mod, 175

Magic numbers, 177
Magnetic moment
   due to spin, 101
   intrinsic, 101
   of atoms, 100–104
   of nuclei, 157
   orbital, 101
Magneton
   Bohr, 101
   nuclear, 157
Maser, 204
Mass-energy relation, 27
Mass separation, 190, 192
Mass spectrograph, 56, 191
Mass spectrum, 56
Mass variation with velocity, 27
Mean square radius, 136, 148
Medical imaging
   emission, 204
   nuclear magnetic resonance, 158
   positron emission tomography, 204
   radiography, 202
   tomography, 202
   use of X-rays, 201–202
Microwaves, 16
Millikan's oil-drop experiment, 52
Models
   of atoms, 69–70, 74, 79
   of electricity, 1–2
   of nuclei, 175–180
   role of, 1

Molecules
   existence of, 2
   formation of, 139
Moseley's law, 107, *see also* X-ray
   spectra
Motion of charged particles
   in crossed field, 50
   in electric field, 44–46
   in magnetic field, 46–49
Multiple scattering, 68
Muonic atom, 150

Neutrino, 170
   evidence for, 170
   properties of, 171
Neutron, 143
Nuclear binding energy, 160, 176
   per nucleon, 161
Nuclear energy levels, 155
   lifetime, 155
   width, 155
Nuclear force, 72, 173–175
   range, 174
   saturation, 174
   spin-dependence, 175
Nuclear magnetic moment, 157
Nuclear magnetic resonance, 158
   imaging, 158
   relaxation times, 158
Nuclear magneton, 157
Nuclear models, 175–180
   compound nucleus, 176
   liquid drop, 175
   optical model, 179
   shell model, 178
Nuclear reactions, 150
   energy balance, 159
   Q-value, 159, *see also* Fission; Fusion;
      Radioactivity
Nuclear scattering, 150, *see also* Cross-
      section; Scattering
Nucleons, 143
Nucleus
   constituents, 143
   density, 74
   density distribution, 145
   magnetic moments, 157
   quadrupole moments, 149
   spin, 145, 157

size, 72, 148
valley of stability, 145

Oil-drop experiment, 52
Optical electron, 96

Pair annihilation, 35
Pair production, 34
Particle accelerators, 183–187
Pauli exclusion principle, 110
Periodic Table
    relation to atomic number, 113
    role of exclusion principle, 112
Photoelectric effect, 18–24
    comparison with Compton scattering, 32
    Einstein's equation for, 21
    failure of classical theory, 19
    quantum yield, 24
    threshold frequency, 19
Photon, 21
    momentum of, 29
    velocity of, 21
Physical constants, 8
Planck's constant, 21
Planck's theory of thermal radiation, 20
Positive ions, 43, 55
Positron, 34
Positron emission tomography, 204
Potential barrier, 132, 154, 169
Principle of superposition, 15, 121
Probability density, 128
Proton, 143

Quality factor, 166
Quantum, 21
Quantum defect, 98
Quantum number, 80
    as label, 111
    orbital, 98, 111
    principal, 80, 111
    spin, 100, 111
    total, 101
Q-value of nuclear reaction, 159

Rad, 166

Radioactivity
    activity, 164
    alpha, 167–169
    beta, 170–171
    decay chain, 163, 165
    decay law, 162, 163–164
    disintegration constant, 162
    half-life, 162
    units, 165
Radiofrequency waves, 16
Rayleigh scattering, 37
Rays, 13
Rectilinear propagation, 14
Relativistic mechanics, 26–28
Rem, 166
Resonance fluorescence, 95
Resonance potential, 92
Resonance radiation, 95
Rest mass, 27
Rest mass energy, 27
Roentgen, 167
Rutherford's model of the atom, 69–71
    deficiencies, 74
Rutherford scattering, 70
    cross-section for, 70, 72, 181
    distance of closest approach, 71
Rydberg constant, 76, 81

Scattering
    elastic, 3, 91, 150
    of billiard balls, 3
    inelastic, 5, 91, 150
    multiple, 68
    Rutherford, 70–72
    small-angle, 68
Schrödinger's equation, 123–139
    time-dependent, 142
    time-independent, 125
Selection rules, 78, 99
Shell, 112
    and Periodic Table, 112
    of electrons in atoms, 112–114
    of nucleons in nuclei, 178–179
Sievert, 166
SI Units, 8
Specific charge, 43
    of electrons, 50–51
    of positive ion, 55–56

Spectra, *See* Atomic spectra; X-ray
  spectra
Spectral lines, 75
Spin angular momentum, 101
Spin-orbit interaction,
  in atoms, 106
Spin quantum number, 100
  of electrons, 100
  of nucleons, 143
  of other particles, 144
Spontaneous emission, 204
Stern-Gerlach experiment, 102–103
Stimulated emission, 204
  power amplification by, 205
Stopping potential, 22
Subshell, 112
Synchrocyclotron, 186
Synchrotron, 186
Synchrotron radiation, 187

Thermal radiation, 16, 20
Thomson model of the atom, 69
Thomson scattering, 37
Tunnel effect
  in alpha-decay, 169
  in metals, 132

Uncertainty Principle, 61
  and determinism, 119
  and measurement, 62
Units, 8, 165

Valence electron, 96
Valency, 43
Velocity selector, 50

Wave
  displacement due to, 120
  intensity of, 14
  longitudinal, 120
  transverse, 120
Wave equation, 123
Wavefront, 13
Wave function
  for electrons in box, 130
  for electrons in periodic potential, 132
  for one-electron atoms, 134
  interpretation, 128
  normalization condition, 128
Wavelength, 16
Wave number, 121
Wave packet, 121, 126
Wave-particle duality, 39, 58
Wave velocity, 121
Weak interaction, 144, 175
Work function, 21

X-ray diffraction
  Bragg reflection, 197
  Debye–Scherrer method, 58, 198
  Laue equations, 195
X-rays, 17
X-ray imaging, *See* Medical imaging
X-ray spectra
  absorption edge, 107
  continuous, 24
  discrete, 106
  from muonic atoms, 150
  Moseley's law, 107

Zeeman effect, 103–104
Zero-point energy, 131